26th International Laser Physics Workshop (LPHYS'17)

AA006825

Journal of Physics: Conference Series Volume 999

Kazan, Russia
17 – 21 July 2017

ISBN: 978-1-5108-6126-8
ISSN: 1742-6588

Printed from e-media with permission by:

Curran Associates, Inc.
57 Morehouse Lane
Red Hook, NY 12571

CURRAN ASSOCIATES INC. proceedings.com

Some format issues inherent in the e-media version may also appear in this print version.

Printed by Curran Associates, Inc. (2018)

For permission requests, please contact the Institute of Physics
at the address below.

Institute of Physics
Dirac House, Temple Back
Bristol BS1 6BE UK

Phone: 44 1 17 929 7481
Fax: 44 1 17 920 0979

techtracking@iop.org

Additional copies of this publication are available from:

Curran Associates, Inc.
57 Morehouse Lane
Red Hook, NY 12571 USA
Phone: 845-758-0400
Fax: 845-758-2633
Email: curran@proceedings.com
Web: www.proceedings.com

Table of contents

Volume 999

26th Annual International Laser Physics Workshop (LPHYS'17)

17–21 July 2017, Kazan, Russian Federation

Accepted papers received: 22 March 2018
Published online: 5 April 2018

Preface

Papers

Modern Trends in Laser Physics

Strong Field & Attosecond Physics

Physics of Lasers

Nonlinear Optics & Spectroscopy

Physics of Cold Trapped Atoms

Quantum Information Science

Fiber Optics

IOP Conf. Series: Journal of Physics: Conf. Series **999** (2018) 011001 doi:10.1088/1742-6596/999/1/011001

EDITORIAL

Dear Readers,

The 26th annual International Laser Physics Workshop, LPHYS'17, took place in the city of Kazan, Russia. 267 participants from 30 countries attended the conference. It was hosted by the hosted by the Kazan E K Zavoisky Physical-Technical Institute (KPhTI).

The LPHYS'17 Steering Committee and the Advisory & Program Committee would like to extend their sincere gratitude to Prof. Alexey A Kalachev (Chair of the Local Organizing Committee) and to his team for the outstanding job performed on organizing, arranging, managing and putting in order the conference. Their combined effort lead to a successful result.

In this volume of the IOP Journal of Physics: Conference Series you will find selected proceedings of the Workshop in Kazan.

Please make a note that the 27th annual International Laser Physics Workshop (LPHYS'18) will take place from July 16 to July 20, 2018 in the city of Nottingham, United Kingdom, at the University of Nottingham, hosted by the Institute of Physics.

With kind regards,

Steering and Advisory & Program committees

LPHYS'17

IOP Conf. Series: Journal of Physics: Conf. Series **999** (2018) 011002 doi:10.1088/1742-6596/999/1/011002

Peer review statement

All papers published in this volume of *Journal of Physics: Conference Series* have been peer reviewed through processes administered by the proceedings Editors. Reviews were conducted by expert referees to the professional and scientific standards expected of a proceedings journal published by IOP Publishing.

Raman Spectroscopy of Novel UHMW Polyethylene-Based Nanocomposites with Nanographite and Nanoclay

K A Prokhorov[1], E A Sagitova[1], A A Averin[2], G Yu Nikolaeva[1], A V Baimova[1,3], L A Novokshonova[4], P N Brevnov[4] and P P Pashinin[1]

[1]Prokhorov General Physics Institute of RAS, 38 Vavilov St., 119991 Moscow, Russia. Contact Phone: +74995038768

[2]Frumkin Institute of Physical Chemistry and Electrochemistry of RAS, 31 Leninsky Prospect, 199071 Moscow, Russia

[3]Moscow Aviation Institute (National Research University), 4 Volokolamskoe Shosse, 125993 Moscow, Russia

[4]Semenov Institute of Chemical Physics of RAS, 4 Kosygin St., 119991 Moscow, Russia

E-mail: sagitova@kapella.gpi.ru

Abstract. We analyze the Raman spectra of nanocomposites based on ultrahigh-molecular-weight polyethylene with nanoclay, thermoexpanded graphite, and reduced graphite oxide fillers. We discuss the potential of Raman spectroscopy for quantitative analysis of the nanocomposite structure, the influence of the fillers on the phase and conformation compositions of the polymer matrix, as well as for the monitoring of dispersion of the nanographite fillers in the nanocomposites.

1. Introduction

Modern industry requires state-of-the-art materials with a set of properties, which neat polymers (so-called homopolymers) in many cases cannot provide. For many applications, this problem can be solved by creation of polymeric blends, composites, and copolymers. For example, ultrahigh-molecular-weight polyethylene (UHMW PE), whose molecular weight M_w exceeds 10^6 Da, possesses the unique set of properties, such as chemical and temperature resistance, the friction coefficient, and physiological inertness [1, 2]. However, enhancement of important performance characteristics, such as stiffness, deformation, heat and wear resistances, is needed [1, 2]. For this reason, composites with UHMW PE matrix and different inorganic fillers are actively developed currently. Nanographite, graphene, and aluminum-silicate nanoclays are the promising fillers, which allow a significant modification of the polymer properties even at low filler content [1-4].

The structural diagnostics of such materials is of great importance, because the properties of any polymer composite material depend strongly on a numerous amount of structural characteristics (conformational and phase compositions of polymer matrix, filler content and sizes, polymer crystallite sizes, the degree of dispersion of a filler in a polymer matrix, etc.) [1-4].

Raman characterization of such objects most fully reveals the unique opportunities of this technique, allowing a nondestructive characterization of all the phases of semi-crystalline polymeric matrix, including non-crystalline areas, in which non-organic filler is localized. In particular, the relative content of crystallites (or the crystallinity degree) and the total content of *trans*-conformers in PE and PE/clay nanocomposites can be evaluated by using the ratio of integral intensity (I^*_{1415}) of the PE Raman band at 1415 cm^{-1} and the reference intensity $I^*_{ref} = I^*_{1295}+I^*_{1305}$, that is the sum of integral intensities of the

PE bands at 1295 and 1305 cm^{-1} [5-9]. The band at 1415 cm^{-1} relates to the wagging vibrations of CH_2 groups in PE macromolecules in the orthorhombic crystalline phase. The bands at 1295 and 1305 cm^{-1} are assigned to the twisting ($\tau (CH_2)$) vibrations of CH_2 groups [5-9]. The former belongs to the vibrations of *trans*-conformers in PE macromolecules, which are localized in both the crystallites and the non-crystalline phase [5]. The latter is the vibrations of PE macromolecules with significant amount of *gauche*-conformers in the non-crystalline phase [5]. The ratio of integral intensities I^*_{1295}/I^*_{ref} is a measure of the total content of *trans*-conformers [5 - 9]. The depolarization ratio of the band at 1295 cm^{-1} gives information about orientational order of macromolecules in PE and PE-based nanocomposites [10, 11].

On the other hand, Raman spectroscopy is powerful technique for structural characterization of nanographite and graphene [12-17]. Analysis of the so-called G, D, 2D bands allows the monitoring of an amount of graphene layers and lateral sizes of graphite nanoplates. The G band, corresponding to sp^2 C=C bonds, is always active in Raman spectra of graphite materials [12-17]. The resonance-enhanced D band is due to structure defects in graphite materials [13-16]. For example, it is active in Raman spectra of micro- and nanographite, but it is not found in the spectrum of graphite monocrystal [14, 15]. The value of lateral size L_a of nanographite is inversely proportional to the ratio I^*_D/I^*_G of integral intensities of the D and G bands [13]. The shape and peak position of the resonance-enhanced 2D-band are reliable fingerprints to identify an amount of layers in graphite materials [17].

Currently, Raman spectroscopy is rarely used for structural study of PE nanocomposite with carbon and aluminum-silicate fillers, because of such composites are relatively novel materials. In particular, there are a few publications, concerning Raman characterization of PE/nanoclay composites [3, 9, 10]. Raman studies [2, 4, 18] of polymer nanocomposites with nanographite (or graphene) fillers are usually limited to structural characterization of the fillers.

In our contribution, we carry out Raman structural study of both polymer matrix and filler for UHMW PE nanocomposites with nanographite and nanoclay. Also, we consider the ways to take into account the contributions from the PE Raman bands, which can overlap with the Raman bands of nanographite. This analysis is required in order to extend the Raman methods of quantitative structural characterization to their applications to PE/graphite nanocomposites.

2. Experimental

We studied the films of the nanocomposites based on UHMW PE matrix ($M_w = 2.0\text{-}2.5 \times 10^6$ Da) with three fillers: Closite 15A (MMT, Southern Clay Products, USA), thermoexpanded graphite (TEG), and reduced graphite oxide (RGO). The composites were synthesized by "*in situ*" polymerization on the catalyst-activated filler surface [1]. The nanocomposite films with thickness of 0.5 mm were obtained via hot compression molding at 160°C and 10 MPa.

The nanocomposites UHMW PE/MMT, UHMW PE/TEG, and UHMW PE/RGO contain 5.3 and 8.3 wt% of MMT, 1.6 and 7.8 wt% of TEG, and 2.0 wt% of RGO, respectively.

The MMT is an industrial Na^+ montmorillonite clay, modified by alkylammonium salts, whose content was about 30 wt%.

In order to produce the graphite nanoplates (nanograpite), the 50% aqueous ethanol solutions of TEG and RGO were treated via UV sonification (200 W, 10 min) with subsequent filtrating and drying at 60°C in vacuum. In accordance with the X-ray diffraction studies [1], the thickness (L_{002}) of the graphite nanoplates was 21 nm that is equal to 80-90 graphene layers. The scanning electron microscopy (SEM) data [1] displayed the lateral sizes of these nanoplates from 1 to 10 μm.

To record Raman spectra of UHMW PE/MMT composites, we used a Sapphire SF 532 laser (Coherent Inc., USA), a double monochromator (U1000, Jobin Yvon, France) and Peltier-element cooled photomultiplier detector, operating in photon counting regime. The excitation wavelength was 532 nm. Spectra were recorded at 90°-scattering with the spectral resolution of 5 cm^{-1}. Diameter of the laser spot on a sample was about 2 mm.

The Raman spectra of the UHMW PE/TEG and UHMW PE/RGO composites, as well as the TEG and graphite powders were recorded by Raman microscope spectrometer "Senterra" (Bruker, Germany).

The excitation wavelengths were 532 and 785 nm. The scattered light was collected in the backscattering geometry using a 20× lens (N.A.=0.65). The laser power at the sample and spectral resolution were 2 mW and 4 cm^{-1}, respectively.

Spectral analysis was performed by the homemade software, which provided the spectra deconvolution via a weighted sum of Gaussian and Lorentzian functions [19] for the PE bands and via Lorentzian fit for the G, D, 2D-bands of the carbon fillers. The baseline was described by a second-order polynomial and was fitted simultaneously with the Raman lines.

3. Results and Discussion

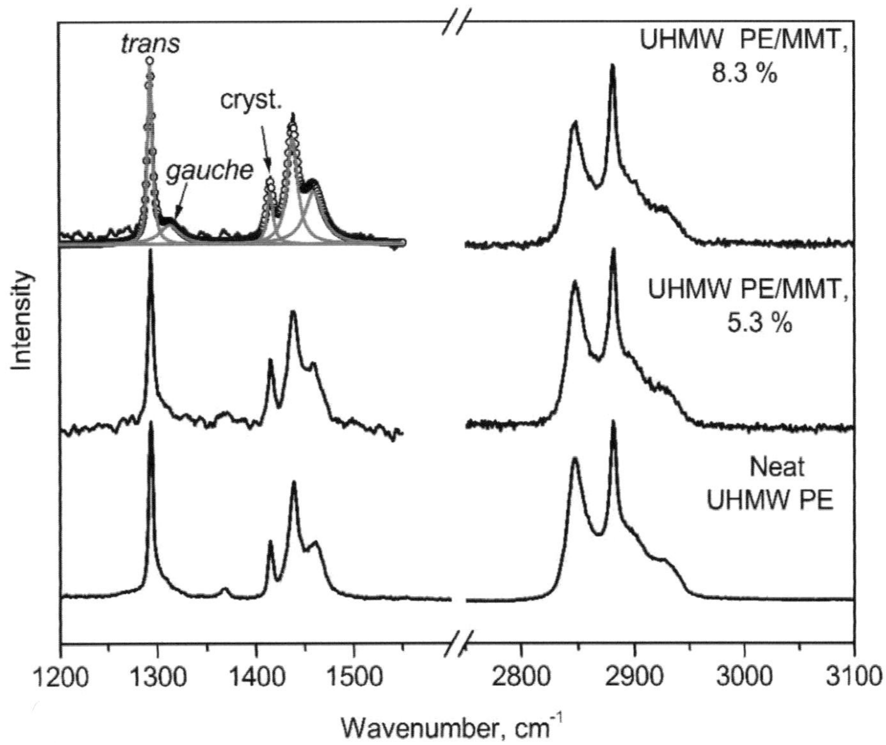

Figure 1. Raman spectra (black) of UHMW PE/MMT composites with 8.3 and 5.3% of MMT contents and neat UHMW PE. The gray lines and open symbols are results of the deconvolution and the simulated spectrum, respectively.

Figure 1 shows the spectra of the neat UHMW PE and UHMW PE/MMT nanocomposites with 5.3 and 8.3 % of the filler contents. For the composite with 8.3 % of MMT, we also have plotted the results of the deconvolution analysis in the spectral region 1200-1550 cm^{-1} as well as the simulated spectrum, which was obtained by summation of the spectral components, found by the deconvolution analysis, and the background.

Figure 2 shows the Raman spectra of the TEG, graphite powder and the UHMW PE film, recorded by using the excitation wavelength $\lambda = 532$ nm, the Raman spectrum of the graphite, recorded by using the wavelength $\lambda = 785$ nm as well as the results of the deconvolution analysis of the 2D band using two lines of Lorentzian/Gauss shapes.

In the spectra, which are plotted in Figures 1 and 2, the intensities in the regions 1200-1700 and 2450-3050 cm^{-1} are normalized to the peak intensity of the bands at 1295 and 2881 cm^{-1}, respectively. Note, that the absolute peak intensity of the band at 2881 cm^{-1} is approximately 4.5 times higher, than that of the band at 1295 cm^{-1}.

3.1. Raman characterization of PE-matrix

The separation of the contributions from the overlapping Raman bands of the fillers and PE-matrix is the main obstacle to quantify the structural characteristics of both the polymer matrix and filler by using Raman spectra.

Raman spectrum of MMT clay, modified by alkylammonium salts, contains the bands of the modifier at 1295 and 1305 cm^{-1} [20, 21]. However, the contributions of these bands to the integral intensities of τ (CH_2)-bands of PE are negligible [10], because the modifier content in polymer composites is usually low (< 3 wt%). Thus, we consider that Raman spectra of the UHMW PE/MMT composites in the regions 1200-1550 cm^{-1} are described by only the PE bands (Figure 1).

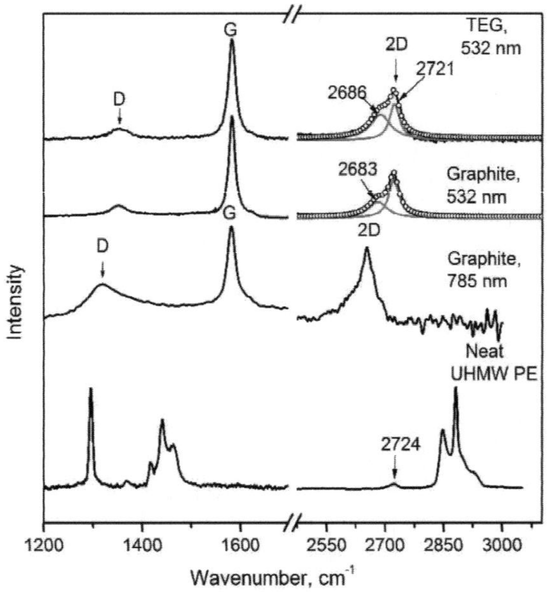

Figure 2. Raman spectra (black) of the TEG, graphite and the neat UHMW PE, recorded at the 532 nm excitation, and Raman spectrum (black) of graphite, recorded at the 785 nm excitation. The gray lines and open symbols are results of the deconvolution analysis and the simulated spectrum, respectively.

We found out that the proper choice of the excitation wavelength allows to solve the problem of "overlapping bands" for PE composites with nanographite (or graphene) fillers. Namely, for the TEG and graphite powders under study the peak position of the D band was found at 1318 cm^{-1} at the excitation wavelength of 785 nm (Figure 2). At the 532 nm excitation the D band peak position is expected to shift to higher wavenumbers (Figure 2). It is observed at 1352 cm^{-1} for the TEG and graphite, and at 1350 cm^{-1} for the composites with the nanographite fillers (Figure 3). It is important to note that the D-band can be wide enough [14, 17]. For our samples at the 532 nm excitation the full width at the half-maximum (FWHM) of the D band was 40-48 cm^{-1} for the composites with the TEG filler and the TEG and graphite powders. FWHM of the D band reaches up 72 cm^{-1} for the nanocomposite with the RGO filler (Figure 3). In spite of such large width of the D band, the deconvolution in the spectral area 1200-1550 cm^{-1} allows to separate the PE components at 1295, 1305, and 1415 cm^{-1} (Figure 3). Thus, the crystallinity degree (α_{cr}) and the total content of *trans*-conformers (α_{trans}) of polymer matrix for the UHMW PE/MMT and UHMW PE/nanographite composites can be estimated from Raman spectra by the relations, suggested by G.R. Strobl and W. Hagedorn for neat PE [5]:

$$\alpha_{cr}(\%) = 100\ \% \times (I^*_{1415}/kI^*_{ref}) \qquad (1)$$

$$\alpha_{trans}(\%) = 100\ \% \times (I^*_{1295}/I^*_{ref}) \qquad (2)$$

Here, the normalization coefficient k=0.43 was calculated from the Raman spectra of solid n-alkanes with the degree of crystallinity of 100% (recorded at the same conditions as the spectra of the neat UHMW PE and the composites).

Table 1 contains the values of α_{cr} and α_{trans}, which were calculated by using the formulas (1) and (2) for the neat UHMW PE and the nanocomposites under study. Also, Table 1 presents the ratio I_{2848}/I_{2881} of the peak intensities of the bands at 2848 and 2881 cm^{-1} (the symmetric and asymmetric stretching vibrations of CH_2 groups, respectively), which decreases at growth of α_{trans} [22].

Table 1. The degree of crystallinity α_{cr}, the total content of *trans*-conformers α_{trans}, and the ratio I_{2848}/I_{2881}, calculated for the neat UHMW PE and for the polymer matrix of the nanocomposites.

Filler content, wt%	α_{cr}, %	α_{trans}, %	I_{2848}/I_{2881}
	UHMW PE/MMT composites		
5.3	38	52	0.81
8.3	40	62	0.68
	UHMW PE/TEG composites		
1.6	40	58	0.73
7.8	53	67	0.70
	UHMW PE/RGO composite		
2.0	41	43	0.81
	Neat UHMW PE		
0	40	54	0.79

The amount of *trans*-conformers in the non-crystalline phase can be estimated as the difference between α_{trans} and α_{cr} [5]. The value (100 % - α_{trans}) can be used as a measure of the content of *gauche* conformers [5].

It can be seen from Table 1 that the non-crystalline phase of the neat UHMW PE and the nanocomposites under study contains mostly PE macromolecules with a numerous content of *gauche* conformers. However, the introduction of the fillers can slightly increase the content of *trans*-conformers in this phase.

3.2. Raman characterization of the nanographite fillers

Figure 3 shows Raman spectra of the UHMW PE/RGO composite with RGO content of 2.0 %, the UHMW PE/TEG composites with TEG contents of 1.6 and 7.8 %, the neat UHMW PE as well as the results of the deconvolution analysis of these spectra in the areas 1200-1550 and 2600-2800 cm^{-1}. The intensities of the bands in the Raman spectra in Figure 3 are normalized to the peak intensity of the G band.

The G band in the spectra of the graphite is observed at 1583 and 1580 cm^{-1} at the excitation wavelengths of 532 and 785 nm, respectively (Figure 2). Also, in the spectra, recorded at $\lambda = 532$ nm, this band is found at 1581 and 1579 cm^{-1} for the UHMW PE/RGO and UHMW PE/TEG composites, respectively (Figure 3). Such shifts of the G band are comparable with the value of the spectral resolution used. However, we observed these shifts for all the spectra recorded at different points of the samples. This result demonstrates that the peak position of the G band is sensitive to both an excitation wavelength and kind of graphitic material. The similar result was observed early for various graphitic materials [14].

We observed the D band for the nanocomposites with 2.0 % of RGO and 7.8 % of TEG filler, but we could not recognize this band in the spectrum of nanocomposite with 1.6 % of TEG filler.

In accordance with the study [13], the lateral size L_a of nanographite plates can be estimated by using the formula:

$$L_a\,(nm) = 2.4\times10^{-10} \times \lambda^4 \times (I^*_D/I^*_G)^{-1} \quad (3)$$

Here, λ (nm) is the excitation wavelength.

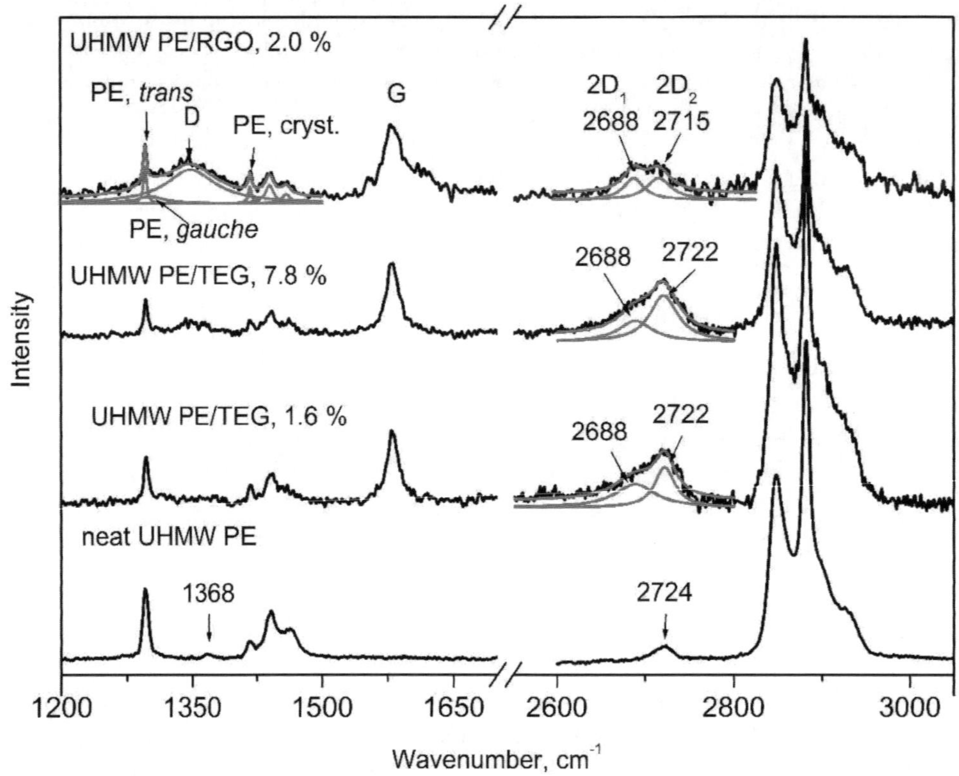

Figure 3. Raman spectra (black) of the UHMW PE/RGO composite with 2.0 % of RGO, the UHMW PE/TEG composites with 1.6 and 7.8 % of TEG, and the neat UHMW PE. The gray lines are results of the deconvolution analysis and the simulated spectrum.

We used the formula (3) and the Raman spectra, recorded at the 532 nm excitation, to evaluate the size L_a for the PE/nanographite composites. We applied two methods of calculations. In the first approach we neglected the contribution from the weak PE band at 1368 cm^{-1} in the intensity I^*_D. In the second approach we took into account the contribution of the PE band at 1368 cm^{-1} as $0.1\times I^*_{ref}$, which was subtracted from I^*_D. The value $0.1\times I^*_{ref}$ is the integral intensity of the PE band at 1368 cm^{-1}, which was calculated by using the Raman spectrum of the neat UHMW PE and was expressed as proportion of the reference intensity I^*_{ref}. The two values calculated by different ways coincided within the range of measurement error.

We estimated the size L_a as 10-11 and 38-40 nm for the UHMW PE/RGO composites with 2.0 % of the filler and for the UHMW PE/TEG composites with 7.8 % of the filler, respectively. Note that these values are much less than the lateral sizes (1-10 μm), which were detected by the SEM study [1]. This fact is explained by different sensitivity of Raman and SEM techniques. In particular, Raman spectroscopy is not effective for evaluation of large lateral sizes (1 μm and more) of particles of graphite materials [14, 15]. Thus, the absence of the D band in the spectrum of the PE/TEG composite with 1.6 % of filler can be a result of relatively large lateral size L_a of filler crystallites.

Note that overlapping of the D and τ (CH_2) bands does not allow to use the excitation wavelength of 785 nm for evaluation of the size L_a of the filler. However, this excitation wavelength is favourable to identify the dispersion degree of filler in PE matrix. At the 785 nm excitation we found out that the 2D band is observed at wavenumber of 2563 cm^{-1}, and it does not overlap with the Raman bands of PE. Unfortunately, the low sensitivity of our spectrometer in IR region did not allow us to record the spectra for the wavenumbers upper than 2740 cm^{-1}.

At the excitation wavelength of 532 nm the 2D band in the Raman spectra of the graphite, TEG, and all the composites under study is of complex shape, which can be described by two contributions that are so – called $2D_1$ and $2D_2$ components [17]. The $2D_1$ component is found at 2683, 2686, and 2688 cm^{-1} for the graphite, the TEG and the composites, respectively. The component $2D_2$ has the peak positions at 2715 cm^{-1} for the UHMW PE/RGO composite, at 2721 cm^{-1} for the graphite, and at 2722 cm^{-1} for the TEG and the UHMW PE/TEG composites.

Decrease in an amount of graphene layers results in growth of intensity of the $2D_1$ component and an increment enhancement of the $2D_2$ component as well as in decrease in the ratio I_{2D}/I_G of intensities of the 2D band and the G band [17]. The intensity of the 2D band is determined as the maximum intensity without spectrum deconvolution. The 2D band splitting vanishes in the Raman spectra of graphene. For instance [17], at the 514.5 nm excitation the Raman spectrum of graphene demonstrates the single 2D band with the peak position at 2688 cm^{-1} and the ratio $I_{2D}/I_G = 5$, while the Raman spectrum of the bilayer structure contains the $2D_1$ and $2D_2$ bands with almost equal intensities and the ratio $I_{2D}/I_G = 2.5$.

However, the Raman spectrum of the neat PE contains a weak band at 2724 cm^{-1}, whose contribution should be taken into account at an analysis of the filler dispersion in the polymer matrix.

For the samples under this study, the values of the ratio I_{2D}/I_G were estimated to be 0.43-0.46 for the graphite, the TEG, and the UHMW PE/RGO composite, while this ratio for the UHMW PE/TEG composites were equal to 0.77 and 0.82 for 1.6 and 7.8 % of filler contents, respectively. For the neat UHMW PE we found out that the peak intensity of the PE Raman band at 2724 cm^{-1} is about 20 % from that of the PE Raman band at 1295 cm^{-1}. In turn, the peak intensity of the PE band at 1295 cm^{-1} in the Raman spectra of all the composites under study is about 50 % from the peak intensity of the G band. In other words, for the investigated composites the contribution of intensity of the PE band at 2724cm^{-1} to the peak intensity at about 2722 cm^{-1} is estimated to be about 10 % from the peak intensity of the G band (we consider that I_G=1). Thus, the observed shift of the $2D_1$ component, the increase in the ratio I_{2D}/I_G in the spectra of the UHMW PE/TEG composites in comparison with the spectrum of the neat TEG is due to the increase in the dispersion degree of the TEG nanoplates in polymer matrix.

However, reliable Raman analysis in the spectral region 2600-2800 cm^{-1} for the UHMW PE/RGO composite needs an additional studies, because of the contradiction between the two experimental observations: the almost equal intensities of the $2D_1$ and $2D_2$ components (indicating presence of the graphene bilayer) and the low intensity ratio $I_{2D}/I_G = 0.43$ (indicating presence of graphite). We consider that this effect is due to a noticeable contribution of the PE band at 2724 cm^{-1} to the intensity of the $2D_2$ component.

4. Conclusions

In this study we analyzed the Raman spectra of nanocomposites of ultrahigh molecular weight polyethylene (UHMW PE) and nanoclay and two grades of UHMW PE/nanographite nanocomposites – with thermoexpanded graphite (TEG) and reduced graphite oxide (RGO) fillers.

Our main conclusions are:

1. The Raman method [5], elaborated for quantitative characterization of the phase and conformational compositions of neat polyethylene (PE), can be applied without any modification to an analysis of the same characteristics of PE matrix in nanocomposites with nanoclays or nanographite filles. It is important to note that in this case spectra of the PE/nanographite composites should be recorded at the excitation wavelength of 532 nm. The applicability of this method [5] for analysis of Raman spectra recorded at other excitation wavelengths is required an additional study, because of overlapping of Raman bands of PE and nanographite.
2. At the excitation wavelength of 532 nm the Raman method [13], elaborated for an estimation of the lateral sizes of nanographite particles, can be extended to PE/nanographite composites without any modification.
3. The analysis of the 2D band in the composite Raman spectra, recorded at the excitation wavelength of 532 nm, allows qualitative characterization of the dispersion degree of the filler in PE/nanographite

composites. However, the excitation wavelength of 785 nm is more preferable for evaluation of the dispersion degree, because of absence of overlapping of the 2D band of the filler and the PE bands.

4. For the UHMW PE/nanoclay and UHMW PE/nanographite composites under study we found that their non-crystalline areas contain mostly PE macromolecules with a numerous content of *gauche* conformers. At the same time, introduction of the fillers can slightly increase the content of *trans*-conformers in these areas.

5. We found out that the TEG nanoplates in the polymer matrix consist of a smaller amount of graphene layers in comparison with the source TEG.

Acknowledgments
This work is supported by the Russian Foundation for Basic Research (project code 16-02-00808-a).

References

[1] Brevnov P N, Kirsankina G R, Zabolotnov A S, Krasheninnikov V G, Grinev V G, Berezkina N G, Sinevich E A, Shcherbina M A and Novokshonova L A 2016 *Polym. Sci.* B **58** 38-49
[2] Bhattacharyya A, Chen S and Zhu M 2014 *eXPRESS Polym. Lett.* **8** 74-84
[3] Guseva M, Gerasin V and Shklyruk B 2016 *J. Appl. Polym. Sci.* 43819
[4] Saner Okan B 2017 *Turk. J. Chem.* **41** 381-90
[5] Strobl G R and Hagedorn W 1978 *J. Polym. Sci.: Polym. Phys. Ed.* **16** 1181–93
[6] Keresztury G and Foldes E 1990 *Polym. Test.* **9** 329–39
[7] Zavgorodnev Yu V *et al* 2012 *Laser Phys.* **22** 730–7
[8] Zavgorodnev Yu V, Prokhorov K A, Nikolaeva G Yu, Sagitova E A, Pashinin P P, Ushakova T M, Novokshonova L A, Starchak E E and Krasheninnikov V G 2013 *Laser Phys.* **23** 025701
[9] Sagitova E A 2008 *Raman Structural Study of Polyethylene—Based Nanocomposites*, PhD Thesis (Moscow: Prokhorov General Physics Institute)
[10] Sagitova E A, Prokhorov K A, Nikolaeva G Yu, Pashinin P P, Gerasin V A, Guseva M A and Antipov E M 2008 *Laser Phys.* **18** 868–81
[11] Nikolaeva G Yu, Prokhorov K A, Pashinin P P and Gordeyev S A 1999 *Laser Phys.* **9** 955–8
[12] Ferrari A C 2007 *Solid State Commun.* **143** 47-57
[13] Cançado L G *et al* 2006 *Appl. Phys. Lett.* **88** 163106
[14] Wang Y, Alsmeyer D C and McCreery R L 1990 *Chem. Mater.* **2** 557-63
[15] Tuinstra F and Koenig J L 1970 *J. Chem. Phys.* **53** 1126-30
[16] Heise H M, Kuckuk R, Ojha A K, Srivastava A, Srivastava V and Asthana B P 2009 *J. Raman Spectrosc.* **40** 344–53
[17] Das A, Chakraborty B and Sood A K 2008 *Bull. Mater. Sci.* **31** 579-84
[18] King A A K, Davies B R, Noorbehesht N, Newman P, Church T L, Harris A T, Razal J M and Minett A I 2016 *Sci. Rep.* **6** 19491
[19] Zavgorodnev Yu V, Chvalun S N, Nikolaeva G Yu, Sagitova E A, Pashinin P P, Gordeyev S A and Prokhorov K A 2015 *J. of Phys.: Conf. Ser.* **594** 012010
[20] Sagitova E A, Donfack P, Prokhorov K A, Nikolaeva G Yu, Gerasin V A, Merekalova N D, Materny A, Antipov E M and Pashinin P P 2009 *J. Phys. Chem.* B **113** 7482–90
[21] Sagitova E A, Donfack P, Prokhorov K A, Nikolaeva G Yu, Gerasin V A, Merekalova N D, Materny A, Antipov E M and Pashinin P P 2012 *J. Phys. Chem.* B **116** 221–31
[22] Tanaka H and Takemura T 1983 *Jpn. J. Appl. Phys.* **22** 1001-4

Raman analysis of polyethylene glycols and polyethylene oxides

E A Sagitova[1], K A Prokhorov[1], G Yu Nikolaeva[1], A V Baimova[1,2], P P Pashinin[1], A Yu Yarysheva[3] and D I Mendeleev[4]

[1] Prokhorov General Physics Institute of RAS, 38 Vavilov St., 119991 Moscow, Russia
[2] Moscow Aviation Institute (National Research University), 4 Volokolamskoe Shosse, 125993 Moscow, Russia
[3] Department of Chemistry, Lomonosov Moscow State University, 1/3 Leninskiye Gory, 119991 Moscow, Russia
[4] Topchiev Institute of Petrochemical Synthesis of RAS, 29 Leninsky Prospect, 119991 Moscow, Russia

Email: sagitova@kapella.gpi.ru

Abstract. We present Raman study of commercial liquids and powders of polyethylene glycols and polyethylene oxides with the average molecular weight from 400 Da to 10000 kDa. The most significant spectral changes were observed for the range of the molecular weights, where the liquid/semisolid transition has occurred. For the powders we revealed increase in the content of the molecules in the helical conformation and in the content of the monoclinic crystalline phase with growth of the molecular weight.

1. Introduction

Polyethylene glycols (PEGs, OH-(CH_2-CH_2-O)$_n$-H) and polyethylene oxides (PEOs, -(CH_2-CH_2-O)$_n$-) are widely used in numerous areas of human activity, in particular, in chemical, food, pharmaceutical and cosmetic industries, in mechanical engineering, metal and oil processing, in agriculture. One of the rapidly growing state-of-the-art technologies is application of PEGs for creation of so-called PEGylated pharmaceuticals, which are used at treatment of various diseases, including some types of cancer and hepatitis, AIDS, and as drug delivery systems. Furthermore, PEGs/PEOs are extensively used in fundamental research, for example, as simple models of biomolecules.

Aggregate state of PEGs/PEOs (liquid, semisolid, or solid), conformational (helical 7_2 or *trans*-conformation) and packing order (monoclinic or triclinic crystalline lattice, the degree of crystallinity), physical and chemical properties, toxicity and photo-/thermal stability and, as consequence, possible areas of applications depend strongly on the molecular weight. For these substances the transition from liquid to semisolid state is observed at molecular weights from 700 to 1000 Da (depending on the purification that is the content of water and impurities/stabilizers). Using differential scanning calorimetry (DSC) and X-ray diffraction analysis Majumdar *et al.* [1] have found that the degree of crystallinity of PEG/PEO powders increases with growth of the molecular weight, reaches the maximum at the molecular weight of 8000-9000 Da, and then decreases slowly (the maximum investigated molecular weight was 23000 Da). Thus, because of numerous vital applications it is very important to be able to characterize the structure and state of PEG/PEO molecules and to identify the molecules of various molecular weights in both the neat forms, and in different compositions and blends, including PEGylated forms.

Furthermore, the PEG/PEO molecules form the homologous series, and these substances are commercially available in a wide range of molecular weights. Studying a homologous series is the way to investigate the supramolecular structure and properties of the series of molecular substances with sequential and stepwise changes in the molecule structure, realized by increase in the number of repeat units (so-called, the homological difference). This way, one can obtain extensive information about the structure - properties relations and to develop models, which allow predicting the supramolecular structure and the properties of uninvestigated homologs. This knowledge opens new ways for development of novel functional materials with desired properties by slight modification of the molecule structure.

There are two stereoregular conformations of PEG/PEO molecules: helical conformation 7_2 (seven repeat units in the *trans-gauche-trans* (TGT) conformation of sequences -O-CH_2-CH_2-O- on two turns in an identity period) and *trans*-conformation. Based on data of X-ray diffraction analysis, the monoclinic crystalline lattice with four molecules in the unit cell was reported for PEG/PEO molecules in the distorted conformation of the helix 7_2 [2] and the triclinic crystalline lattice with one molecule in the unit cell – for the molecules in the *trans*-conformation [3]. The triclinic modification is observed in stretched PEOs and is stable only under tension.

Raman spectroscopy is a highly informative, non-destructive and label-free technique of molecular spectroscopy, which can be effectively applied for both liquids and solids. In particular, using Raman spectra it is possible to evaluate the configurational order, the conformational composition (the contents of conformationally ordered monomer sequences of a certain length), the packing order (the degree of crystallinity, the type of the crystalline lattice(s), and the distribution of crystallite sizes), the chemical composition and the content of various chemical groups, the orientational order of molecules. However, Raman spectroscopy has been rarely used for systematic and detailed study of the homologous series of PEG/PEO molecules. Probably, it is connected with complex Raman spectra of these substances, where overlapping of numerous lines of various sequences of conformers is observed. This also makes the assignment of the PEG/PEO Raman lines not straightforward, and there are many contradictions between the assignments and peak positions of the Raman lines even in a few Raman studies of PEGs/PEOs, which were found in literature [4-12].

Based on experimental Raman study of PEG with the molecular weight of 6000 Da and the normal coordinate analysis on a large number of conformers of model molecules, Matsuura and Fukuhara [4,5] have suggested assignment of the Raman lines of PEGs/PEOs in the spectral region from 800 to 1500 cm^{-1} to the vibrations of various sequences of conformers. In particular, it was shown that the spectra of various conformer sequences are strongly overlapping, and it is a reason of difficulties in Raman analysis of conformational composition of PEG/PEO molecules. Temperature-dependent evolution of Raman spectra of PEG with the molecular weight of 2000 Da was studied in detail in the spectral range about 800 cm^{-1} [6].

Raman spectroscopy was applied to study the conformation order of PEG macromolecules of the molecular weight of 1500 Da [7,8]. The lines at 1227 and 1275 cm^{-1} were assigned to the CH_2 wagging vibrations of *trans*- and *gauche*-conformers relative to C-C bond, respectively, while the lines at 1118 and 1136 cm^{-1} - to the CH_2 twisting vibrations of *trans*- and *gauche*-conformers relative to C-O bond, respectively. Analyzing changes in integral intensities of these lines, Kozielski *et al.* [7,8] supposed that with decrease in PEG concentration in water solutions the ratio of contents of the PEG molecules in the helical and *trans* conformations increases. Using IR spectroscopy, similar results were obtained for a number of short-chain PEGs with various terminal groups in water solutions with varying PEG concentration [9]. Based on their proposed assignments of the Raman lines at 1118, 1136, 1227, and 1275 cm^{-1}, Kozielski *et al.* [7] have stated that the conformational order of PEG/PEO molecules is practically independent on the molecular weight in the range from 1500 to 12000 Da.

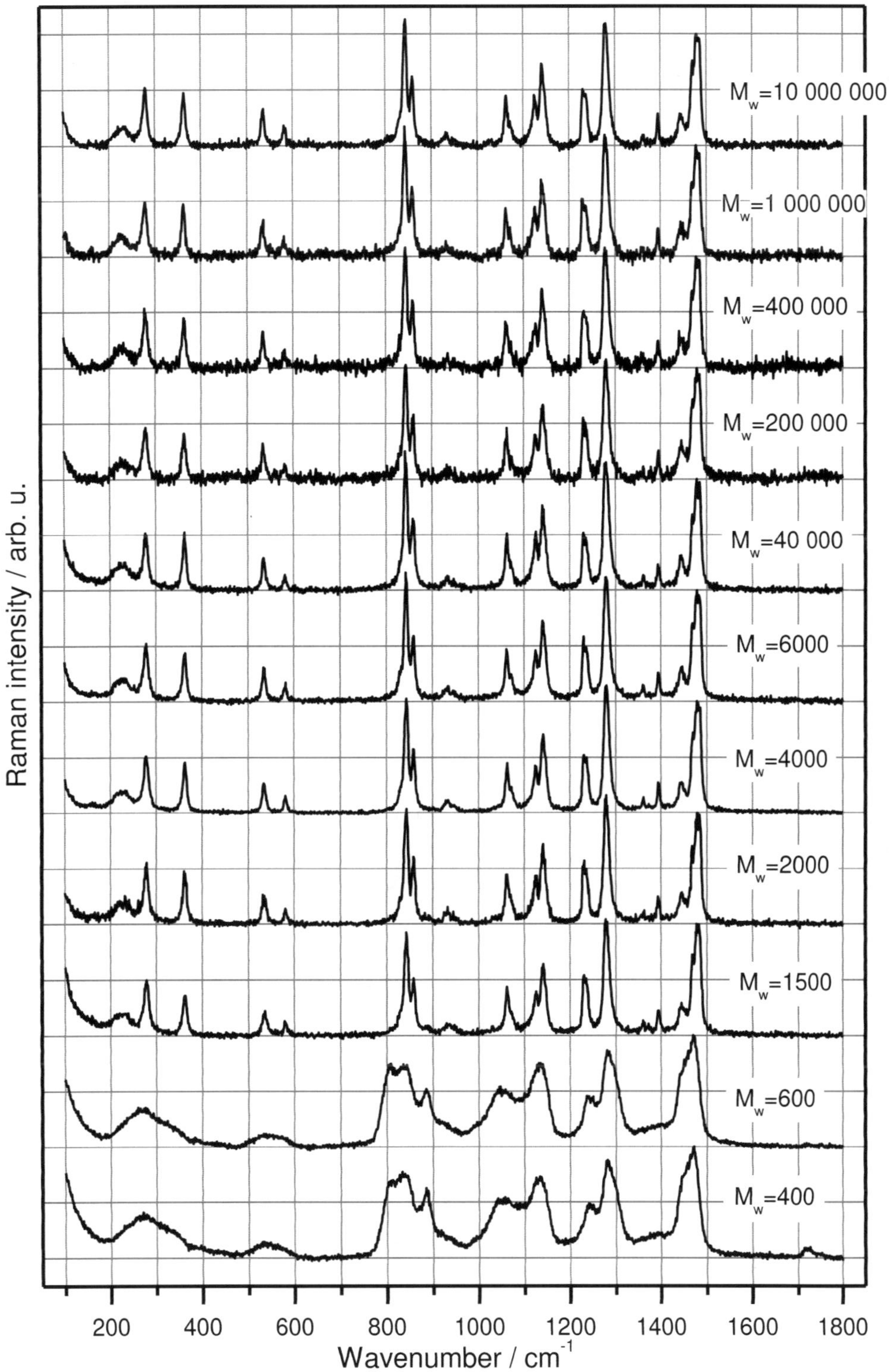

Figure 1. Raman spectra of PEGs/PEOs in the region from 100 to 1800 cm^{-1}. M_w is the average molecular weight in Daltons. PEGs with the molecular weight of 400 and 600 Da are liquids, while the other samples are semisolids and solids.

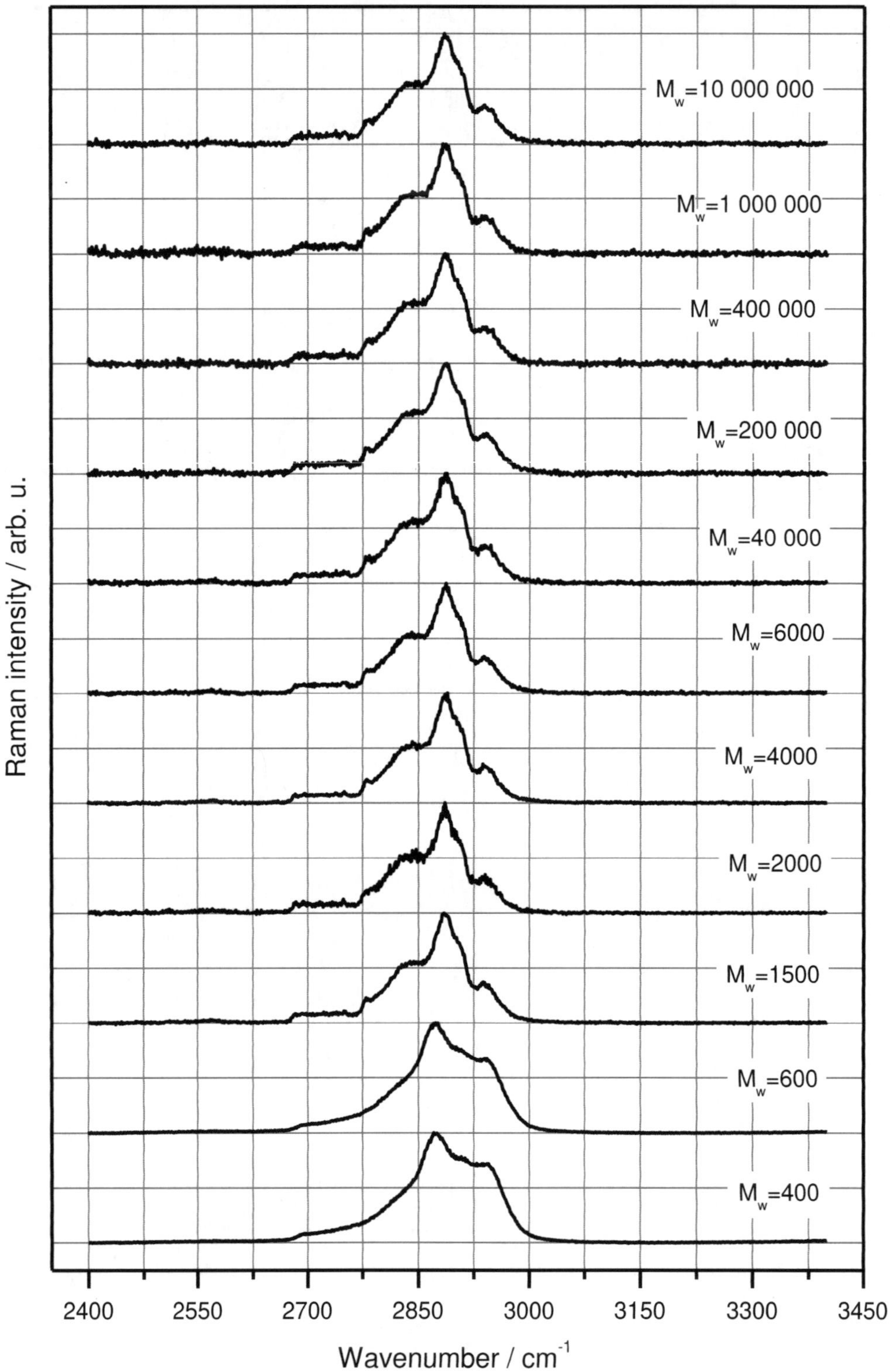

Figure 2. Raman spectra of PEGs/PEOs in the region from 2400 to 3400 cm^{-1}. M_w is the average molecular weight in Daltons. PEGs with the molecular weight of 400 and 600 Da are liquids, while the other samples are semisolids and solids.

However, the assignment of Kozielski *et al.* [7,8] is not supported by results of Matsuura and Fukuhara [4,5]. On the other hand, the lines at 1124 and 1280 cm^{-1} were related to the vibrations of PEG/PEO molecules in the helical conformation TGT in the monoclinic crystalline phase [10].

Samuel and Umapathy [10] carried out temperature-dependent Raman study of PEG of the molecular weight of 1000 Da as well as the DFT calculations of the Raman spectrum of molecule CH_3-O-(CH_2-CH_2-O)$_6$-CH_3, which was chosen as a model of one repeat unit of helix 7_2.

Yang *et al.* [11] studied Raman spectra of PEO with the molecular weight of 4600 Da in aqueous solution and in the melt. Using the normal coordinate calculations, Raman spectra of PEO were simulated by superposing the calculated spectra of an ensemble of various conformers.

Yang *et al.* [12] have investigated the effect of molecular weight on PEG/PEO molecular conformation (helix or *trans*-) by X-ray diffraction analysis, FTIR, and DSC. It was found that in the case of films, prepared by the solution-casting method, growth of the molecular weight leads to the decrease in the PEG/PEO crystallinity and the content of molecules in the helical conformation with simultaneous increase in the content of molecules in the *trans*-conformation.

Summing up, assignments of the Raman lines of PEGs/PEOs are still incomplete, and the qualitative and quantitative relationships between the Raman spectra and the molecular and supramolecular structure are not yet fully known. In this work we present Raman spectroscopic study of the series of commercial PEGs/PEOs with average molecular weight, varied in a wide range.

2. Experimental

PEG/PEO samples with the average molecular weight from 400 Da to 10000 kDa were purchased from BASF и FERAK Berlin companies. The samples with the molecular weights of 400 and 600 Da were liquids, while the other samples were in the forms of semisolids or powders. In this work the samples were analyzed as they were purchased, that is without any processing or treatment.

The Raman setup consisted of a Sapphire SF 532 laser (Coherent Inc., USA), a double monochromator U1000 (Jobin Yvon, France) and Peltier-element cooled photomultiplier detector, operating in the photon counting regime. The excitation wavelength was 532 nm. Non-polarized Raman spectra were recorded at room temperature at 90°-scattering with the spectral resolution of 5 cm^{-1}. Diameter of the laser spot on a sample was about 2 mm.

Analysis of the experimental spectra was carried out using the homemade software. Shape of each line was described by the weighted sum of Lorentz and Gauss functions. Baseline was described as a second-order polynomial and fitted simultaneously with the Raman lines.

3. Results and Discussion

Figures 1 and 2 present the Raman spectra of PEGs/PEOs with the average molecular weights from 400 Da to 10000 kDa in the region from 100 to 1800 cm^{-1} and in the region from 2400 to 3400 cm^{-1}, respectively. The spectra in the region from 100 to 1800 cm^{-1} are normalized to the maximum Raman intensity in the range 1450-1500 cm^{-1}, where the lines, assigned to the CH_2 bending vibrations [4], are manifested. This maximum intensity is observed at about 1470 cm^{-1} for the liquid PEGs and at about 1480 cm^{-1} for the semisolid and solid PEGs/PEOs. The spectra in the region from 2400 to 3400 cm^{-1} are normalized to the peak intensity of the most intense line in each spectrum. The Raman-active lines of PEGs/PEOs are absent in the region from 1800 to 2400 cm^{-1}.

We found out that the peak positions and intensities of the PEG/PEO Raman lines are sensitive to the change in the molecular weight (Figures 1, 2, and 3). Most significant spectral changes with the increase in the molecular weight are observed for PEGs with the molecular weights of 600 and 1500 Da that is at the transition from liquid to semisolid state for our limited set of the samples. At the same time, the Raman spectra of two liquid PEGs (400 and 600 Da) are very similar as well as the Raman spectra of all the semisolid and solid PEGs/PEOs do not change dramatically with the increase in the molecular weight.

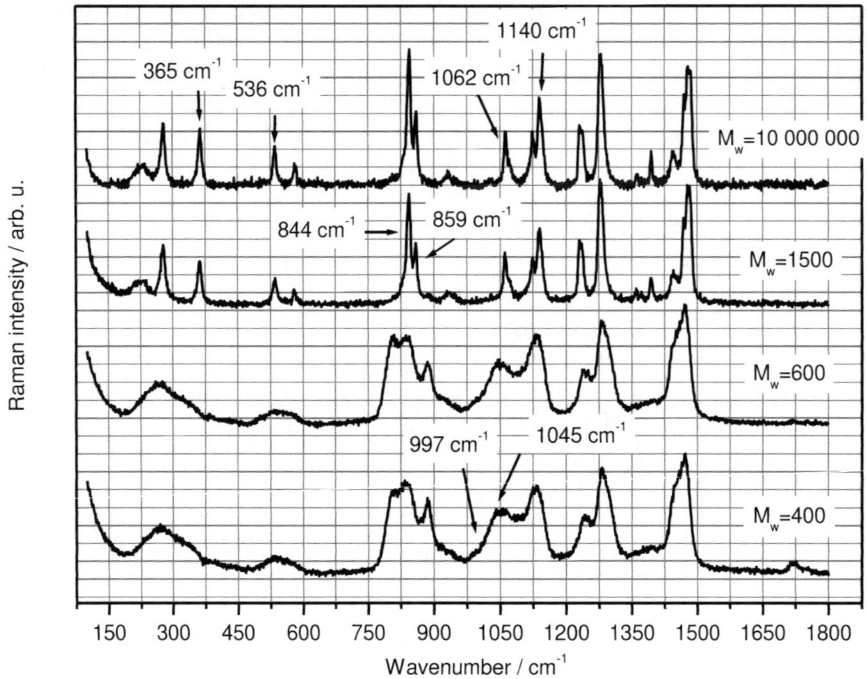

Figure 3. Raman spectra of PEGs/PEOs in the region from 100 to 1800 cm^{-1}. M_w is the average molecular weight in Daltons.

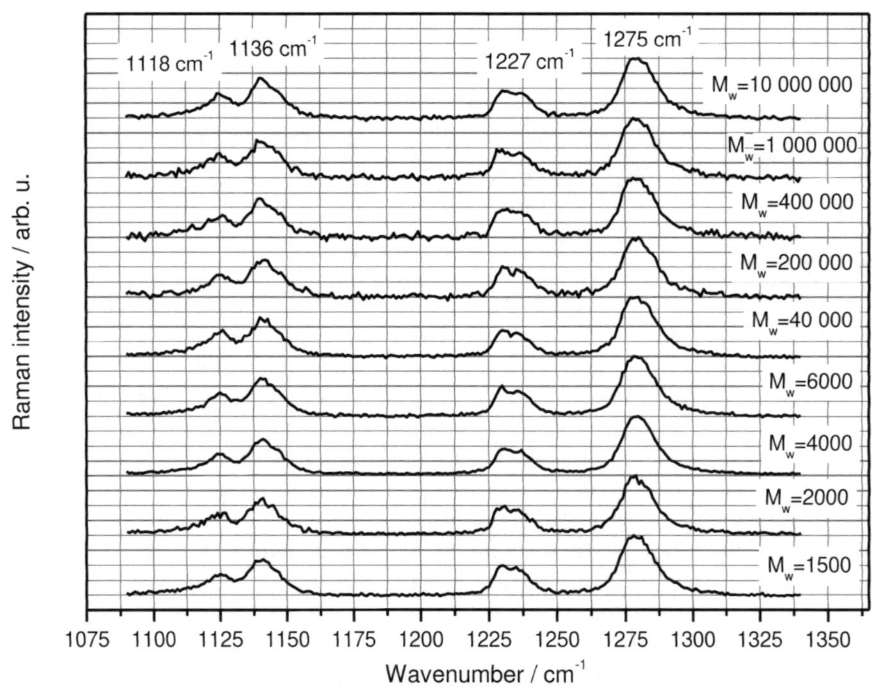

Figure 4. Raman spectra of the semisolid/solid PEGs/PEOs in the region from 1090 to 1340 cm^{-1}. M_w is the average molecular weight in Daltons. The spectra are normalized to the peak intensity of the line at about 1280 cm^{-1}.

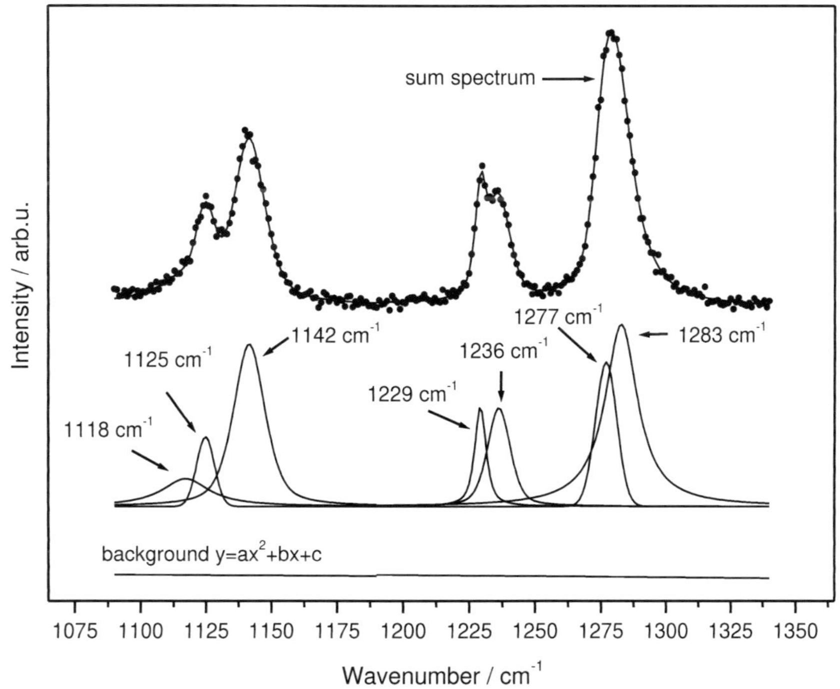

Figure 5. Deconvolution of the Raman spectrum of PEG with the molecular weight of 6000 Da in the region from 1090 to 1340 cm^{-1}: dots – experimental data, lines – simulated spectra.

Let's first compare the spectra of two liquid PEGs under study. We found out three noticeable spectral distinctions, that are, first, presence of the weak line at about 1720 cm^{-1}, second, higher intensity of the line at 880 cm^{-1}, and, third, slightly higher intensity of the line at 2940 cm^{-1} in the spectrum of PEG with the molecular weight of 400 Da compared with the spectrum of PEG with the molecular weight of 600 Da (Figures 1, 2, and 3).

The line at 1720 cm^{-1} most probably corresponds to the vibrations of C=O bonds in the formiates and other esters, which were formed due to PEG oxidation in air [13].

Matsuura and Fukuhara [4,5] have assigned the line at 880 cm^{-1} to the superposition of the C-O stretching vibration and the CH_2 rocking vibration of sequences of conformers, distinct from the helical 7_2 structure. These authors [4,5] have observed the line at about 885 - 900 cm^{-1} in the Raman spectra of the melt and aqueous solutions of PEG with the molecular weight of 6000 Da, while this line was absent in the spectrum of the solid state of this PEG. This line is also absent in our spectra of all the semisolid and solid PEGs under study (Figures 1 and 3). In principle, the line at 880 cm^{-1} can be also related [5] to the C-OH stretching vibration of terminal hydroxyl groups in low molecular weight PEGs.

Samuel and Umapathy [10] have noted that the intensities of the Raman lines at 884 and 2940 cm^{-1} of PEG with the molecular weight of 1000 Da increase with temperature growth. Kozielski *et al.* [6] have also observed appearance of the band at 880 cm^{-1} in Raman spectra of PEG with the molecular weight of 2000 Da with heating.

Summing up, we concluded that dumping of the lines at 884 and 2940 cm^{-1} is the evidence that conformational order of PEG molecules with the molecular weight of 600 Da is higher compared with that of PEG with the molecular weight of 400 Da.

The second step of our analysis is comparison of the spectra of semisolid and solid PEGs/PEOs with each other. In spite of the weak spectral variations for this series of the samples, we observed slight, but reliable increase in the intensities of the lines at 365, 536, 844, 859, and 1140 cm^{-1} with growth of the molecular weight (Figures 1 and 3). In the region from 2400 to 3400 cm^{-1} the Raman spectra of the semisolid and solid PEGs are very similar (Figure 2).

Yang *et al.* have assigned [11] the line at 365 cm^{-1} to the C-O-C bending vibration of PEG/PEO molecules in the helical conformation. This band relates to the disordered longitudinal acoustic mode (D-LAM), corresponding to straight chain segments with a well defined specific chain conformation. This line is absent in the spectra of the liquid PEGs (Figures 1 and 3). The peak position of this line is the same for all the semisolid/solid PEGs/PEOs under study (within the limits of the spectral resolution used).

Samuel and Umapathy [10] have shown that heating of PEG of the molecular weight of 1000 Da leads to damping of the lines at 844 (the C-O stretching vibration + the CH_2 rocking vibration), 863 (the C-C stretching vibration + the C-O stretching vibration + the CH_2 rocking vibration), and 1140 cm^{-1} (the C-O stretching vibration + the CH_2 rocking vibration) in its Raman spectrum. These authors [10] have assigned the line at 844 cm^{-1} to the vibrations of molecules in the helical conformation in the monoclinic crystalline phase. They showed that intensity of the line at 844 cm^{-1} decreases at heating and increases at cooling, and these variations are in very good agreement with DSC thermograms.

Based on experimental and calculated Raman spectra, Yang *et al.* [11] have assigned the band at 848 cm^{-1} to the C-O stretching vibrations of molecules in the helical conformation. Kozielsky *et al.* [6] have observed the decrease in intensity of the lines at 838 and 855 cm^{-1} in Raman spectra of PEG with the molecular weight of 2000 Da with temperature growth. Matsuura and Fukuhara [5] have found that the calculated PEG/PEO Raman spectra contain the lines in the range 825-870 cm^{-1}, related to the sequences of conformers xGx-xGx (where x implies either T or G) in chain segments -O-CH_2-CH_2-O-CH_2-CH_2-O-.

Yang *et al.* [12] have related the IR line at 1140 cm^{-1} to the vibrations of molecules in the monoclinic crystalline phase.

Thus, we concluded that the increase in the intensities of the lines at 365, 844, and 1140 cm^{-1} with a growth of the PEG/PEO molecular weight (Figure 3) should be interpreted as increase in the content of molecules in the helical conformation and in the content of the monoclinic crystalline phase.

Figure 4 presents Raman spectra of the semisolid and solid PEGs/PEOs in the region from 1090 to 1340 cm^{-1}. The spectra are normalized to the peak intensity of the line at about 1280 cm^{-1}. In this region the lines at 1118, 1136, 1227, and 1275 cm^{-1} should be observed. As it was mentioned above, these lines were assigned to *trans-* and *gauche-*conformers of PEG/PEO molecules relative to C-O and C-C bonds [7,8]. It can be seen from Figure 4, that the distribution of the intensity of Raman scattering in this spectral range is not changed with growth of the molecular weight for all the semisolid and solid PEGs/PEOs under study.

Our deconvolution analysis (Figure 5) has shown that the structure of the PEG/PEO Raman spectra in this range is more complex, than it was proposed in [7,8]. In the spectra of the semisolid and solid PEGs/PEOs there are at least seven Raman-active lines with the peak positions at 1118, 1125, 1142, 1229, 1236, 1277, and 1283 cm^{-1}. At the first stage of our deconvolution analysis we supposed presence of only five lines at 1125, 1142, 1229, 1236, and 1280 cm^{-1}, which can be easily recognized in the PEG/PEO spectra. However, the spectra were poorly described by five lines, and, thus, two additional lines were introduced.

Matsuura and Fukuhara [4,5] have found six lines in this region in the spectra of solid PEG, that are the lines at 1112, 1127, 1143, 1233, 1239, and 1282 cm^{-1}. However, these authors did not carry out the deconvolution analysis of the spectra, and, thus, two strongly overlapped lines at 1277 and 1283 cm^{-1} could not be resolved. The lines at 1233 and 1239 cm^{-1} were related to the TGT conformers in chain segments -O-CH_2-CH_2-O-, which form the helical conformation of PEG/PEO molecules [4,5].

Thus, we conclude that the assignment of the PEG/PEO Raman lines in the region from 1090 to 1340 cm^{-1}, suggested by Kozielski *et al.* [7,8], requires additional justifications.

Finally, we proceed to the analysis of the spectral alterations at the liquid/semisolid transition, which in our limited set of the samples is observed at increase in the PEG molecular weight from 600 to 1500 Da. First, we observed the abrupt high-wavenumber shift (from 2874 to 2887 cm^{-1}) of the peak position of the symmetric C-H stretching vibration. Exactly the same effect was observed by Samuel and Umapathy [10] at heating of PEG with the molecular weight of 1000 Da, that is the shift of the line at 2887 cm^{-1} to 2874 cm^{-1}.

The second feature at the transition from the liquid to semisolid PEG state is the sharp decrease in the intensity of the band at 810 cm^{-1}, which was assigned [10] to the vibrations of the sequences of conformers GGT, TGG, and GGG of chain segments -O-CH_2-CH_2-O-. However, Matsuura and Fukuhara [5] have calculated the lines in the region 810-825 cm^{-1}, arising due to conformer sequences xTx (x denotes either G or T) of chain segments -O-CH_2-CH_2-O-.

Samuel and Umapathy [10] showed that the line at 810 cm^{-1} becomes more intense with temperature growth in Raman spectra of PEG with the molecular weight of 1000 Da. Thus, according to their assignment, the conformational order in the liquid PEGs (as well as in the melt of PEG with the molecular weight of 1000 Da [10]) differ from the conformational order of the semisolid/solid PEGs/PEOs in higher content of *gauche*-conformers about C-O bonds, while preference of *gauche*-conformation about C-C bonds is observed for both liquid and semisolid/solid states (as well as for the PEG melt [10]). In [6] appearance of the line at about 800 cm^{-1} was observed at heating of PEG with the molecular weight of 2000 Da.

Samuel and Umapathy have stated [10] that the ratio of intensities of the lines at 810 and 844 cm^{-1} is a measure of relative contents of *gauche*- and *trans*-conformers relative to C-O bonds. The reduction in intensity of the line at 844 cm^{-1} with heating corresponded very well with the increase in intensity of the line at 810 cm^{-1}, indicating the transformation of one type of conformers to another [10]. However, it is not convenient and accurate to calculate the conformational composition of PEG/PEO molecules using these two lines due to strong overlapping of numerous lines in this spectral range. Samuel and Umapathy [10] have found at least 9 lines in this spectral region.

Third, a number of new Raman lines appear in the spectra of the semisolid/solid PEGs/PEOs compared with the liquids. Besides the line at 365 cm^{-1}, the most prominent new features are the lines at about 932, 1362, and 1391 cm^{-1}. Two lines at 930 and 934 cm^{-1} were observed only in the spectra of solid PEG and were assigned to the superposition of the CH_2 rocking, the C-O stretching, and the C-C stretching vibrations of PEG/PEO molecules in the helical conformation [4]. This assignment was confirmed by calculation of the lines of the conformer sequences xGT-TGx of chain segments -O-CH_2-CH_2-O-CH_2-CH_2-O- in the region 940-950 cm^{-1} [5]. The assignment of this feature to the vibrations of molecules in the amorphous domains of solid PEGs/PEOs [10] seems to be in error.

According to Matsuura and Fukuhara [4], the lines at 1364 and 1397 cm^{-1} are related to the CH_2 wagging vibration for most of possible PEG/PEO conformations. However, the line at 1364 cm^{-1} was not observed in Raman spectra of the melt and aqueous solutions of PEG with the molecular weight of 6000 Da [4]. Most probably this line corresponds to the calculated vibrations of the conformer sequences xGT-TGx of chain segments -O-CH_2-CH_2-O-CH_2-CH_2-O- in the region 1350-1360 cm^{-1} [5].

At the same time, Matsuura and Fukuhara [4,5] have noted that the Raman bands at 997 and 1045 cm^{-1} are characteristic of the PEG/PEO melts and aqueous solutions and can correspond to a number of possible conformations of PEG/PEO molecules. Line at 997 cm^{-1} was related [5] to the vibrations of the conformer sequences, containing *trans*-conformers relative to C-C bond and *gauche*-conformer relative to C-O bond.

Indeed, the spectra of semisolid and solid PEGs/PEOs (Figures 1 and 3) do not contain Raman scattering at 997 and 1045 cm^{-1}, but the narrow and strong line at 1060 cm^{-1} is evident. The spectra of the liquid PEGs under study contain broad band with complex structure and the peak position at about 1045 cm^{-1}.

4. Conclusions

We studied Raman spectra of the homological series of polyethylene glycols and polyethylene oxides, that are liquid, semisolid, and solid homologs for the range of the molecular weight from 400 Da to 10000 kDa. Most prominent spectral alterations were observed for the range of the molecular weight, where the liquid/semisolid transition has occurred. At the same time, the spectra of the liquids were very similar as well as the changes in the spectra of the semisolids/solids with growth of the molecular weight were weak. The Raman lines at 810, 880, 997, and 1045 cm^{-1} are characteristics of liquid polyethylene glycols and are absent (or very weak) in the spectra of semisolid/solid polyethylene glycols and polyethylene oxides. These lines were also observed by other authors only in the melts and aqueous solutions of polyethylene glycols and polyethylene oxides.

We found that the lines at 365, 932, 1362, and 1391 cm^{-1} are manifested only in the spectra of semisolids and solids. Based on the increase in the intensity of the lines at 365, 844, and 1140 cm^{-1}, we concluded that the content of molecules of polyethylene glycols and polyethylene oxides in the helical conformation and the content of the monoclinic crystalline phase increase with growth of the molecular weight of these molecules.

The obtained results can be useful for development of new materials for various applications as well as for Raman characterization of molecular substances with similar chemical structure. However, additional investigations of polyethylene glycols and polyethylene oxides in liquid, semisolid, and solid states as well as in the melt and solutions are required for development of Raman methods of precise and reliable diagnostics of structure of these substances. In particular, the assignments of the most of the Raman lines of polyethylene glycols and polyethylene oxides as well as quantitative relationships between the Raman spectra and the molecular and supramolecular structure should be clarified in detail.

Acknowledgements

This work is supported by the Russian Foundation for Basic Research (project code 16-02-00808-a).

References

[1] Majumdar R, Alexander K S and Riga A T 2010 *Pharmazie* **65** 343-7
[2] Takahashi Y and Tadokoro H 1973 *Macromolecules* **6** 672-5
[3] Takahashi Y, Sumita I and Tadokoro H 1973 *J. Polym. Sci.* B **11** 2113-22
[4] Matsuura H and Fukuhara K 1985 *J. Mol. Struct.* **126** 251-60
[5] Matsuura H and Fukuhara K 1986 *J. Polym. Sci.* B **24** 1383-1400
[6] Kozielski M, Muhle M and Blaszczak Z 2004 *J. Mol. Liq.* **111** 1-5
[7] Kozielski M, Muhle M, Blaszczak Z and Szybowicz M 2005 *Cryst. Res. Technol.* **40** 466-70
[8] Kozielski M 2006 *J. Mol. Liq.* **128** 105-7
[9] Begum R and Matsuura H 1997 *J. Chem. Soc., Faraday Trans.* **93** 3839-48
[10] Samuel A Z and Umapathy S 2014 *Polym. J.* **46** 330-6
[11] Yang X, Su Z, Wu D, Hsu S L and Stidham H D 1997 *Macromolecules* **30** 3796-802
[12] Yang S, Liu Z, Liu Y and Jiao Y 2015 *J. Mater. Sci.* **50** 1544-52
[13] Han S, Kim C and Kwon D 1997 *Polymer* **38** 317-23

IOP Conf. Series: Journal of Physics: Conf. Series **999** (2018) 012003 doi:10.1088/1742-6596/999/1/012003

Theoretical treatment of the processes involving the dipole transitions to the lowest exciton states in hexagonal semiconductors

L E Semenova

Prokhorov General Physics Institute, Russian Academy of Sciences, Vavilov St. 38, 119991, Moscow, Russia

E-mail: sl@kapella.gpi.ru

Abstract. The treatment of the two-photon transitions to the $A_{n=1}$ exciton level and the resonant Raman scattering of light by LO-phonons is given for the hexagonal semiconductors A^2B^6, taking into account the influence of the complex top valence band and anisotropy of the exciton effective mass.

1. Introduction

In the semiconductor crystals of the wurtzite structure the top valence band is known to be split into three closely-lying subbands. The transitions from them to the lowest conduction band lead to the excitonic hydrogen-like series A, B and C in the absorption spectra [1,2]. And the energy difference between the upper subbands is less than the exciton Rydberg for some crystals [2]. Taking into account the influence of anisotropy and the complicated structure of the top valence band by the perturbation theory, for the hexagonal semiconductors A^2B^6 the wave functions and the energies in a first approximation were reported to be obtained for the $A_{n=1}$ and $B_{n=1}$ excitonic levels under assumption of the zero wave vector of an exciton [3]. In this work the first order corrections to the energy and the wave functions of the $A_{n=1}$ exciton level with the small nonzero wave vector $\boldsymbol{K}$ were derived. Taking them into account, the treatment of the two-photon dipole transitions to the $A_{n=1}$ level and the resonant Raman scattering of light by LO-phonons is given for the hexagonal semiconductors A^2B^6.

2. Theoretical model

The exciton with the wave vector $\boldsymbol{K}$ is assumed to be formed by an electron with the degenerate conduction band c and by a hole from the degenerate (or almost generate) valence band v [4,5]. The pertinent wave functions of the relative electron-hole motion $\Phi^{(\lambda)}_{j_e j_h}(\boldsymbol{r};\boldsymbol{K})$ are supposed to satisfy the system of equations of the effective mass approximation (EMA) [4-6]. Here and henceforth the indices j_e and j_h denote the states in the conduction band c and the valence band v, respectively. In the case of the hexagonal semiconductors A^2B^6 one can introduce the wave function $\tilde{\chi}^{(\lambda)}_{j_e j_h}(\boldsymbol{r})$ for which the system of the EMA equations is reduced to the form [5]

$$\left[\frac{1}{2\mu_{j_e j_h}^{\perp}}\left(\hat{p}_x^2+\hat{p}_y^2\right)+\frac{1}{2\mu_{j_e j_h}^{||}}\hat{p}_z^2+U_{eh}(\boldsymbol{r})\right]\tilde{\chi}_{j_e j_h}^{(\lambda)}(\boldsymbol{r})+\sum_{j_h'}\left(\delta_{j_h' j_h}-1\right)\hat{V}_{j_h' j_h}\tilde{\chi}_{j_e j_h'}^{(\lambda)}(\boldsymbol{r})=\left(E_\lambda(\boldsymbol{K})-\tilde{\varepsilon}_{j_e j_h}(\boldsymbol{K})\right)\tilde{\chi}_{j_e j_h}^{(\lambda)}(\boldsymbol{r}) \quad (1)$$

where $\hat{p}_\alpha=-i\hbar\,\mathrm{d}/\mathrm{d}r_\alpha$, $\left(\mu_{j_e j_h}\right)^{-1}=\left(m_{j_e}\right)^{-1}+\left(m_{j_h}\right)^{-1}$, m_{j_e} and m_{j_h} are the effective masses of an electron and a hole, $U_{eh}(\boldsymbol{r})$ is the Coulomb energy of the electron-hole interaction, $E_\lambda(\boldsymbol{K})$ is the exciton level energy. $\tilde{\varepsilon}_{j_e j_h}(\boldsymbol{K})$ has the form

$$\tilde{\varepsilon}_{j_e j_h}(\boldsymbol{K})=\varepsilon_{j_e j_h}+\frac{\hbar^2\left(K_x^2+K_y^2\right)}{2\left(m_{j_e}^{\perp}+m_{j_h}^{\perp}\right)}+\frac{\hbar^2 K_z^2}{2\left(m_{j_e}^{||}+m_{j_h}^{||}\right)} \quad (2)$$

where $\varepsilon_{j_e j_h}$ is the energy gap. Here it is assumed that Ka is considerably smaller than unit where a is the exciton radius. The operator $\hat{V}_{j_h' j_h}$ is given by

$$\hat{V}_{j_h' j_h}=D_{j_h' j_h}^{\beta\alpha}\left(\hat{p}_\alpha-\hbar K_\alpha\frac{m_{j_h'}^{\alpha\alpha}}{m_{j_e}^{\alpha\alpha}+m_{j_h'}^{\alpha\alpha}}\right)\left(\hat{p}_\beta-\hbar K_\beta\frac{m_{j_h'}^{\beta\beta}}{m_{j_e}^{\beta\beta}+m_{j_h'}^{\beta\beta}}\right) \quad (3)$$

where $D_{jj'}^{\alpha\beta}$ are the effective mass parameters [4]. For the wurtzite crystals the nonzero parameters $D_{jj'}^{\alpha\beta}$ can be found due to the known zone structure and the symmetry selection rules. Neglect of the coupling between the subbans ($\hat{V}_{j_h' j_h}=0$) and the influence of anisotropy can reduce the equation (1) to the hydrogen-like equation for which the wave functions $\chi_{j_e j_h}^{(\xi\ell m)}(\boldsymbol{r})$ are known [7]. Here $(\xi\ell m)$ is the proper set of quantum numbers for the relative electron-hole motion.

Taking into account the terms associating the equations for the different subbands ($\hat{V}_{j_h' j_h}$, $j_h'\neq j_h$) and anisotropy of the excitonic effective mass $\mu_{j_e j_h}$ by the perturbation theory, the wave functions and the energies in a first approximation were obtained for the $A_{n=1}$ exciton level. The first order corrections to the energy are independent on the wave vector $\boldsymbol{K}$ and lead to the level shift.

The obtained first order corrections to the wave functions $\sum_{\xi,\ell,m} b_{j_e j_h}^{(\xi\ell m)}\chi_{j_e j_h}^{(\xi\ell m)}(\boldsymbol{r})$ involve some nonzero coefficients $b_{j_e j_h}^{(\xi 1 m)}$, as distinct from the case K=0. This can in particular lead to the added sequences of the intermediate exciton states in the processes of the two-photon absorption and the Raman scattering (RS). The expressions for the nonzero coefficients $b_{j_e v_i}^{(\xi\ell m)}$ are too cumbersome and are not presented here.

3. Results and discussion

It is well known that the two-photon dipole transition to the exciton state of the *s*-type can be described within the framework of the two-band as well as three-band model [8,9]. In the first case absorption of a photon is accompanied by the weakly-forbidden dipole transition from the ground state to an exciton of the *p*-type, then the intraband dipole transition to the *s*-exciton occurs when another photon is absorbed. The sequence of the intermediate excitonic states *p-s* corresponds to this process. In accordance with the three-band model the intermediate states are assumed to be the *s*-excitons belonging to the some higher-lying conduction band and the top valence band or to the lowest conduction band and the deeper-lying valence band. However the inclusion of the first order corrections to the exciton wave functions can lead to the added two-photon transition mechanism which is described by following sequence of processes: the allowed dipole transition to the *s*-exciton of the B or C series and the dipole transition to the $A_{n=1}$ level. Thus this process is described within the framework of the two-band model and includes the allowed dipole transitions. Moreover the contributions of the two-band and three-band models are opposite in sign [8,10]. In consequence of this, it was possible to assume that the role of this process can be non-negligible.

In order to compare the added mechanism with the others, the two-photon transition to the $A_{n=1}$ level was considered for the two monochromatic light beams incident on a CdS crystal. The first beam with the frequency ω_1 and polarization $\boldsymbol{\varepsilon}_1 \parallel z$ propagates along the x axis and the second one with ω_2 and $\boldsymbol{\varepsilon}_2 \parallel x$ travels along the z axis. The frequencies ω_1 and ω_2 are assumed be such that their sum $\omega_1 + \omega_2$ lies in the region of resonance with the $A_{n=1}$ level. Consideration of the two-photon transitions was carried out for the near frequencies ω_1 and ω_2. In the case of the unperturbed wave functions under simultaneous absorption of the two photons $\hbar\omega_1$ and $\hbar\omega_2$ the transition is allowed to the exciton state of the symmetry Γ_5 and is described within the framework of the two-band and three-band models. Rough estimations performed have shown that the contribution of the three band model dominates. In this work comparison between the two-photon transition process, which is due to the first order corrections to the exciton wave functions, and the mechanism described by the sequence of the intermediate excitonic states *p-s* of the A series was performed. The terms including the transitions between the nearest bands were assumed to give the main contribution to the parameters $D^{\alpha\beta}_{vv'}$. The estimation of the parameter of the weakly-forbidden transition M_{cv} from Ref. [11] and the data on the effective masses of an electron and a hole from Ref. [12] were used. The computations were performed for the following values of the exciton Rydberg and the energy gap of the A, B and C series: $R_A = R_B = 0.028\,eV$, $R_C = 0.026\,eV$ [1], $E^A_{cv} = 2.578\,eV$, $E^B_{cv} = 2.593\,eV$, $E^C_{cv} = 2.656\,eV$ [12]. Thus the rough estimates obtained for CdS have shown that the contribution of the added process is considerably less than the one of the sequence *p-s*.

The treatment of the Raman scattering of light by LO-phonons under excitation near the $A_{n=1}$ exciton level is of interest too. The sequences of the intermediate excitonic states, *s-p* and *s-s*, are known to give a main contribution to the resonant RS [11,13]. In the first case, the intraband Frohlich exciton-lattice interaction is considered in the approximation of the zero wave vector of a phonon (q=0), but the scattering process includes the weakly-forbidden dipole transitions. The RS mechanism described by the sequence *s-s* is caused by the forbidden q-dependent Frohlich interaction [13].

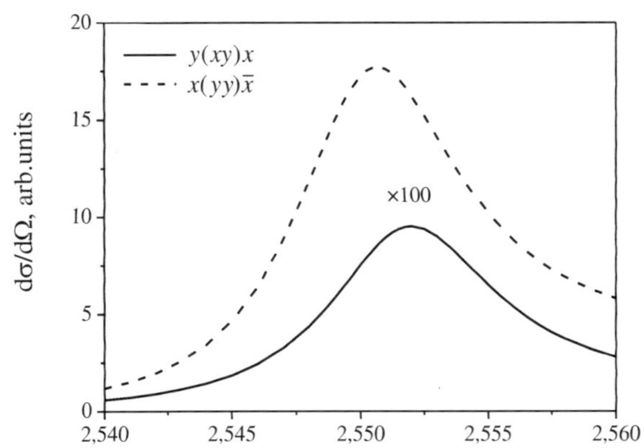

Figure 1. The RS cross section calculated for the scattering mechanism including the dipole transition to the $A_{n=1}$ exciton level in a CdS crystal as a function of the energy of incident photons. The curve dσ/dΩ corresponding to the scattering geometry $y(xy)x$ is multiplied by the factor of 100.

If the first order corrections to the exciton wave functions are taken into account, the added scattering process, which is induced by the q-independent Frohlich interaction (q=0) and comprises only the allowed dipole transitions, can participate in the RS. Its contribution in the scattering cross section was compared with the one of the sequence of the intermediate states *s-s* for a CdS crystal. For this purpose the frequency dependences of the RS cross section dσ/dΩ were computed for the scattering mechanism, which includes dipole transition to the $A_{n=1}$ exciton level and the indirect transition to the ground state and dominates in the considered resonant conditions. Calculations were carried out for the two scattering configurations $x(yy)\bar{x}$ and *y(xy)x*. The results obtained for the lifetime broadening of the exciton level $\Gamma = 0.004\,eV$ [2] are presented in Fig. 1. In the back scattering configuration $x(yy)\bar{x}$ the Raman scattering is described by the sequence of the intermediate states *s-s*. And in this case the cross section was calculated for the unperturbed wave functions. For the geometry *y(xy)x* the resonant RS is only due to inclusion of the corrections to the exciton wave functions because the sequences of the intermediate states *s-s* and *s-p* do not participate in it. Owing to this the frequency dependence of $d\sigma/d\Omega$ was computed for the added scattering process described above. As can be seen from Fig. 1 the RS cross section calculated with due regard for the first order corrections is considerably less than the one obtained for the *s-s* sequence.

4. Conclusion

Taking into account the influence of the complex top valence band and anisotropy of the exciton effective mass by the perturbation theory, the wave functions and the energies in the first approximation were derived for the $A_{n=1}$ exciton level with the nonzero wave vector. The two-photon dipole transitions to the $A_{n=1}$ exciton level and the resonant Raman scattering of light by LO-phonons were considered. The inclusion of the first-order corrections to the wave functions was found to lead to the added sequences of the intermediate states in these processes. Rough estimations obtained for a CdS crystal have shown that their contributions are considerably less than the ones which were calculated for the unperturbed wave functions.

Acknowledgments

This study was partially supported by the Program III.2 of the Russian Academy of Sciences “Fundamental problems of laser technologies”.

References

[1] Reynolds D C 1969 Excitons in II-VI compounds *Optical Properties of Solids* ed S Nudelman and S S Mitra (New York: Plenum Press) chapter 10 pp 239-86
[2] Sobolev V V 1980 *Bands and Excitons of the $A^{II}B^{VI}$ Group Compounds* (Chisinau: Shtiintsa)
[3] Semenova L E, Nikolaeva G Yu and K A Prokhorov 2017 *J. Phys.: Conf. Ser.* **826** 012004
[4] Denisov M M and Makarov V P 1973 *Phys. Stat. Sol.* (b) **56** 9-59
[5] Makarov V P 1971 *Phys. Stat. Sol.* (b) **44** 475-88
[6] Dresselhaus G 1956 *J. Phys. Chem. Solids* **1** 14-22
[7] Landau L D and Lishitz E M 1989 *Course of Theoretical Physics* vol 3 (Moscow:Nauka)
[8] Rustagi K C, Pradere F and Mysyrowicz A 1973 *Phys. Rev.* B **8** 2721-32
[9] Mahan G D 1968 *Phys. Rev.* **170** 825-38
[10] Semenova L E and Prokhorov K A 2003 *JETP* **96** 922-31
[11] Semenova L E and Prokhorov K A 2001 *Bull. Lebedev Phys. Inst.* **2** 39-43
[12] Nguyen B A, Nguyen V H, Nguyen T T and Nguyen A V 1982 *Phys. Rev.* B **25** 4075-80
[13] Martin R M 1971 Phys. Rev. B 4 3676-85

Experimental Evidence for Wigner's Tunneling Time

N Camus [1], E Yakaboylu[1,2], L Fechner[1], M Klaiber[1], M Laux[1], Y Mi[1], K Z Hatsagortsyan[1], T Pfeifer[1], C H Keitel[1], and R Moshammer[1]

1 Max Planck Institute for Nuclear Physics, Saupfercheckweg, 1, Heidelberg, Germany
2 IST Austria, Am Campus 1, Klosterneuburg, Austria

E-mail: nicolas.camus@mpi-hd.mpd.de

Abstract. Tunneling of a particle through a barrier is one of the counter-intuitive properties of quantum mechanical motion. Thanks to advances in the generation of strong laser fields, new opportunities to dynamically investigate this process have been developed. In the so-called attoclock measurements the electron's properties after tunneling are mapped on its emission direction. We investigate the tunneling dynamics and achieve a high sensitivity thanks to two refinements of the attoclock principle. Using near-IR wavelength we place firmly the ionization process in the tunneling regime. Furthermore, we compare the electron momentum distributions of two atomic species of slightly different atomic potentials (argon and krypton) being ionized under absolutely identical conditions. Experimentally, using a reaction microscope, we succeed in measuring the 3D electron momentum distributions for both targets simultaneously. Theoretically, the time resolved description of tunneling in strong-field ionization is studied using the leading quantum-mechanical Wigner treatment. A detailed analysis of the most probable photoelectron emission for Ar and Kr allows testing the theoretical models and a sensitive check of the electron initial conditions at the tunnel exit. The agreement between experiment and theory provides a clear evidence for a non-zero tunneling time delay and a non-vanishing longitudinal momentum at this point.

1. Introduction

Modern laser technology can provide electric fields comparable to those electrons experience in atoms [1], opening opportunities to investigate the quantum-dynamical process of electron tunneling through the potential barrier formed by the superposition of both laser and atomic fields [2] (Fig. 1 a). Attosecond-time [3] and Ångström-space resolution of the strong laser-field technique [4] allow to address fundamental mechanisms related to tunneling, which are still open and debated[5, 6, 7, 8, 9]: What time is spent under the barrier and what momentum is picked up by the particle in the meantime? While these questions address the fundamental question how the electron transition occurs from the (quantum) bound state to the (quasi classically detected) continuum via the tunneling through the barrier, it has also relevant applications for attosecond spectroscopy because common techniques such as high-harmonic generation [10, 11] or laser-induced electron diffraction [12, 13] rely also on tunnel ionization.

To specifically probe the tunneling step one takes advantage of the fact that the electron, once ionized, is driven by the laser electric field. Using close-to-circularly polarized laser pulses, and therefore a rotating electric field, the instant of time when the electron appears in the continuum is effectively mapped onto a characteristic emission direction that can be measured after the pulse (attoclock configuration) [3, 14, 15, 6]. However, in order to gain a meaningful

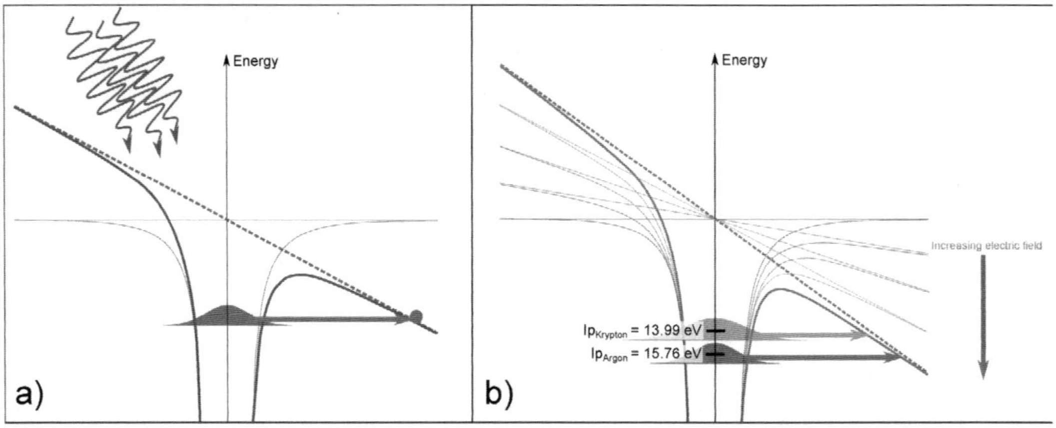

Figure 1. a) Schematic representation of the tunneling ionization: the Coulomb potential is bent by the laser electric field forming a barrier in the potential. The bound electron wave packet is ionized through tunneling and considered to appear as a classical particle at the tunnel exit. b)Scheme to attain a high sensitivity on the barrier influence in tunneling ionization: comparing two atomic species with similar ionization potential.

interpretation of the tunneling step (time, momentum, and position right after tunneling) by inspection of the final photoelectron momentum distribution, the laser pulse parameters must be known and the Coulomb interaction with the remaining ion along the electron excursion in the continuum must be taken into account. Moreover, the Coulomb field effect depends itself on the initial conditions and cannot be evaluated independently. These considerations lead to an ongoing debate about the role of the initial electron momentum [16, 5, 17, 18, 19] as well as the question whether and how the electron motion can be traced back to the tunneling step [15, 6, 8].

Our strategy to achieve a high sensitivity on the tunneling step is to compare strong-field tunneling ionization of atomic species with slightly different ionization potentials (argon and krypton), but under otherwise absolutely identical laser fields (Fig. 1 b). We therefore effectively reduce the influences of other effects such as differences in Stark shifts, atom polarizabilities, dependences on the initial-state orbital momentum [16], and systematic experimental errors, as well as reduce the effect of the Coulomb field. Similarly to other attoclock investigations, we vary in addition the laser intensity such that we can investigate a large range of effective tunneling barrier widths.

Experimentally, we simultaneously collect within the same experiment photoelectron momentum distributions with high resolution for ionization of a gas mixture containing argon and krypton. Using a Reaction Microscope spectrometer that allows for the coincident detection of electron-ion pairs created by ionization of Ar or Kr atoms,we can compare the spectra obtained otherwise under the same laser and experimental conditions. The experimental results allow for a stringent test of theoretical predictions. Our theoretical model is based on the assumption that the electron dynamics is quasiclassical when the electron distance from the tunnel exit exceeds a few de-Broglie wavelengths. The tunneling dynamics is described quantum mechanically via numerical solution of the Schrödinger equation, assuming that the barrier is quasistatic during the brief time of the tunneling. The important point is the matching of the quantum mechanical and classical dynamics. The latter is accomplished employing a quantum trajectory method, which is a generalization of the Wigner concept of the scattering time delay [20]. The quantum trajectory or the Wigner trajectory is derived from the exact wave function. It starts from an under the barrier point and further propagates in the continuum, coinciding with a classical trajectory slightly away from the tunnel exit, and provides the initial conditions for the specific classical trajectory in the continuum. Further classical dynamics in the laser and Coulomb fields yields the asymptotic photoelectron momentum. In this way we deduce the signatures of the tunneling step from the asymptotic photoelectron momentum distribution.

2. Experimental setup

2.1. Reaction Microscope spectrometer

To study the electron momentum distributions for ionization of two atomic systems (argon and krypton) with slightly different ionization potentials I_p within the same experiment we make use of the coincidence capability of the reaction microscope (Fig. 2). By doing such we can avoid the intrinsic experimental inaccuracies of independent measurements (changes in the laser properties and systematic uncertainties).

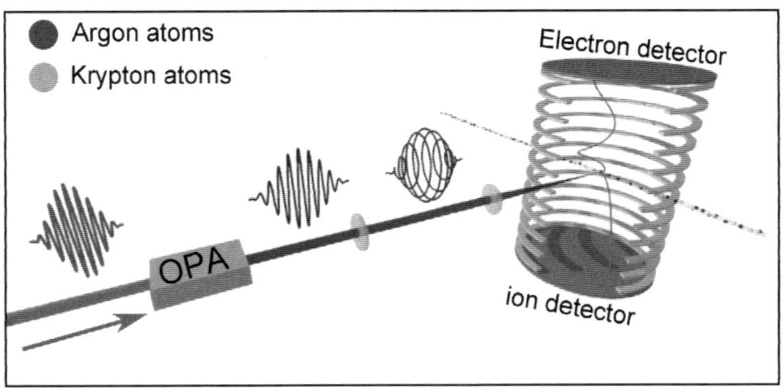

Figure 2. Experimental setup: close to circularly polarized laser pulses with a 1300 nm wavelength are focused onto a gas mixture of argon and krypton inside a reaction microscope.

Linearly polarized 1300 nm radiation was generated by an optical parametric amplification (OPA) system (TOPAS-C, Light Conversion) pumped by 25 fs, 790nm pulses from a Ti:Sapphire based laser system. The ellipticity, using broadband waveplates, was determined to be 0.85 ± 0.05 (the ratio of the minor to the major axis). The laser intensity at the target was adjusted using a motorized iris. The laser pulses are focused onto a supersonic, internally cold gas jet inside a Reaction Microscope (ReMi). The reaction fragments (electrons and ions) are guided by weak homogeneous electric and magnetic fields onto time and position sensitive detectors. From their times of flight and their impact positions, the initial momentum vectors of the fragments are determined [21, 22].

2.2. Observables

In Fig. 3a) the measured electron momentum distributions for argon and krypton for a laser intensity of about $2 \times 10^{14} W/cm^2$ are shown. As expected for atoms with similar ionization potentials, the spectra for both targets appear almost identical. However, clear differences manifest when looking at the difference spectrum of the normalized distributions (Fig. 3a) right panel).

As developed in the theory section below, the value of interest is the most probable electron trajectory, in particular the final momentum and the angle of rotation with respect to the axes of the polarization ellipse [14]. This correspond to the angle θ_{max} of the distribution where the number of counts is maximum. θ_{max} is the observable that is most sensitive to both time delay and initial momentum. In the present case we focus on the angle difference $\Delta\theta_{max} = \theta^{Ar}_{max} - \theta^{Kr}_{max}$ as a function of intensity.

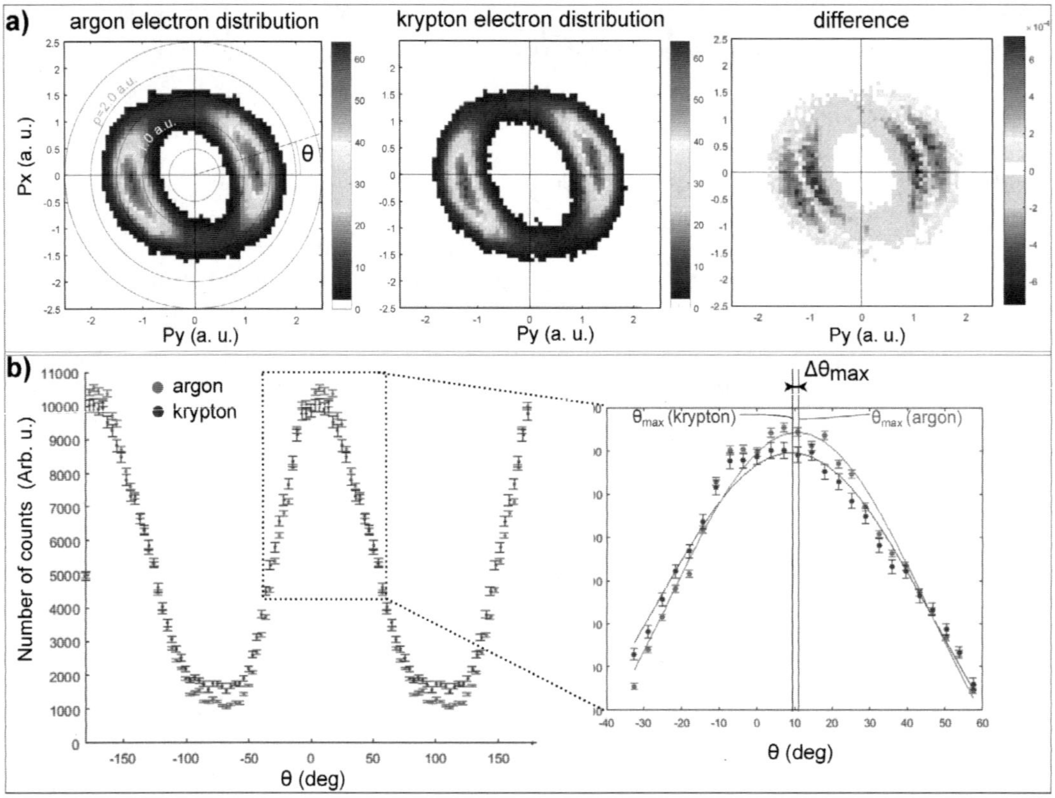

Figure 3. a) Momentum distribution of argon (left), krypton (middle) and the difference (right). b) Angular distribution for argon (red) and krypton (blue). Adapted from [23].

In a strong-field experiment it is almost impossible to measure with high accuracy the absolute value of the intensity since the interaction point is placed in a close vacuum chamber. Instead, the intensity is usually calibrated by comparing the photoelectron and photoion distributions with theory [24]. In the case of the attoclock experiment, the conversion between intensity and radius of the 2D momentum distribution is actually debated [25, 26]. To prevent from any bias, we choose to use the average electron momentum at θ_{max} for ionization of argon: $\rho(\theta_{max}) = \rho^{Ar}(\theta_{max}) = \sqrt{p^{Ar}_x(\theta_{max})^2 + p^{Ar}_y(\theta_{max})^2}$. For circularly and elliptically polarized pulses this average momentum $\rho(\theta_{max})$ is an unambiguous measure of the actual intensity.

The resulting curve $\Delta\theta_{max}$ as a function of $\rho(\theta_{max})$ is plotted in comparison with our theoretical model in Fig. 7.

3. Theoretical model

We develop a model where the tunneling ionization step is treated in the Wigner formalism and the results provided by these calculations are used as initial conditions for a classical propagation of the electron in the continuum.

IOP Conf. Series: Journal of Physics: Conf. Series **999** (2018) 012004 doi:10.1088/1742-6596/999/1/012004

3.1. Wigner trajectory

The quantum mechanical description of the propagation of the ionized electron is given by the propagation of the wave function

$$\Psi(\boldsymbol{r},t) = \langle \boldsymbol{r}|U(t,t_s)|\Psi(t_s)\rangle = \int d\boldsymbol{r}_i\, K(\boldsymbol{r},\boldsymbol{r}_i;t,t_s)\Psi(\boldsymbol{r}_i,t_s)\,, \tag{1}$$

where $\Psi(\boldsymbol{r}_i,t_s)$ is the initial wave function, and the term $K(\boldsymbol{r},\boldsymbol{r}_i;t,t_s) = \langle \boldsymbol{r}|U(t,t_s)|\boldsymbol{r}_i\rangle$ is the corresponding space-time propagator which connects space points $\boldsymbol{r}_i$ and $\boldsymbol{r}$ in a time interval $t-t_s$.

The adiabaticity of the ionization process is guaranteed by a small Keldysh parameter($\gamma < 1$) for most of the parameter regime under consideration here. We can therefore assume that the energy is conserved during the process, and accordingly the space-time propagator can be written in terms of the fixed-energy propagator $G(\boldsymbol{r},\boldsymbol{r}_i,\varepsilon)$, the retarded Green's function of the Time Independent Schrödinger Equation (TISE). After mapping the three-dimensional geometry onto a one-dimensional tunneling geometry in parabolic coordinates [5], the space-time propagator is written as

$$K(\eta,\eta_i,t-t_s) = \frac{1}{2\pi}\int_{-\infty}^{\infty} d\varepsilon\, e^{-i\varepsilon(t-t_s)}G(\eta,\eta_i,\varepsilon)\,, \tag{2}$$

where the Green's function is given by

$$\left(\frac{\varepsilon}{4} + \frac{1}{2}\frac{\partial^2}{\partial\eta^2} - V_B(\eta)\right) G(\eta,\eta_i,\varepsilon) = \delta(\eta-\eta_i)\,. \tag{3}$$

Decomposing the Green's function into its phase and amplitude, by means of the stationary phase analysis we deduce that solutions of the equation

$$(t-t_s) - \frac{\partial \arg\left[G(\eta,\eta_i,\varepsilon)\right]}{\partial\varepsilon} = 0 \tag{4}$$

yield the dominant energies. In other respect, as in the case of tunnel-ionization, if the ionization energy is the dominant energy then the trajectory

$$t(\eta,\eta_i) = \left.\frac{\partial \arg\left[G(\eta,\eta_i,\varepsilon)\right]}{\partial\varepsilon}\right|_{\varepsilon=-I_p} + t_s \tag{5}$$

is the dominant trajectory for the space-time propagator.

We name it Wigner trajectory as it generalizes the Wigner approach for the scattering time delay [20] to the tunneling problem. In the original idea of Wigner the asymptotic time delay at scattering was described by the derivative of the scattering phase shift with respect to energy. The definition of the dominant trajectory is quite natural. In fact, in the WKB approximation, the phase corresponds to the classical action, hence the definition of a trajectory by Eq. (5) would follow the Hamilton-Jacobi theory (see for instance Ref. [27]). In the quantum domain the Wigner trajectory corresponds to the maximum of the electron probability current density [28].

In the case of a one-dimensional problem the Green's function can be calculated:

$$G(\eta,\eta_i,\varepsilon) = \frac{2i}{W}\left[\theta(\eta-\eta_i)\psi_+(\eta,\varepsilon)\psi_-(\eta_i,\varepsilon) + \theta(\eta_i-\eta)\psi_-(\eta,\varepsilon)\psi_+(\eta_i,\varepsilon)\right] \tag{6}$$

with $W = \psi_-(\eta,\varepsilon)\partial_\eta\psi_+(\eta,\varepsilon) - \psi_+(\eta,\varepsilon)\partial_\eta\psi_-(\eta,\varepsilon)$ being the Wronskian [29, 30, 31], where $\psi_+(\eta,\varepsilon)$, and $\psi_-(\eta,\varepsilon)$ are the corresponding solutions of the TISE

$$\left(-\frac{1}{2}\frac{d^2}{d\eta^2} + V_B(\eta)\right)\psi_\pm(\eta,\varepsilon) = \frac{\varepsilon}{4}\psi_\pm(\eta,\varepsilon) \tag{7}$$

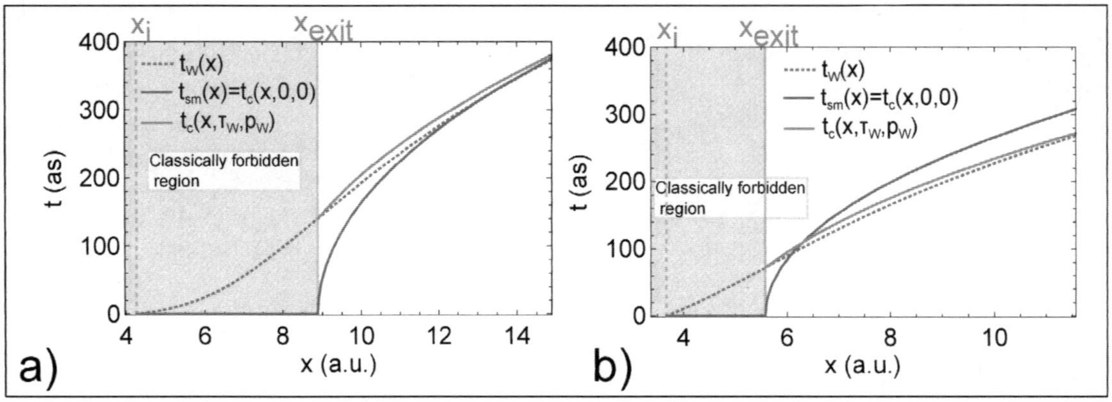

Figure 4. Electron trajectories in different models at low (a) and high (b) laser intensities: (dashed, orange) Wigner trajectory via Eq. (5), (green, solid) the simple-man trajectory, (blue, solid) classical trajectory with the initial conditions given by the Wigner formalism. The trajectories are calculated for a krypton atom. The laser intensities are $I = 1.7$ (a) and $I = 6.1 \times 10^{14}$ W/cm^2 (b). The trajectories are shown in one-dimensional (along the tunneling direction) Cartesian coordinate. Adapted from [23].

with positive and negative current, respectively. The phase of the Green's function is written as

$$\arg\left[G(\eta, \eta_i, \varepsilon)\right] = S_+(\eta, \varepsilon) - S_+(\eta_i, \varepsilon), \quad \eta \geq \eta_i, \tag{8}$$

where $S_\pm \equiv \arg(\psi_\pm)$, and we use the fact that $S_-(\eta_i, \varepsilon) = -S_+(\eta_i, \varepsilon)$. The Wigner trajectory is then defined by the phase of the TISE wave function with a positive outgoing probability current

$$t_W(\eta, \eta_i) = \frac{\partial S(\eta, -I_p)}{\partial \varepsilon} - \frac{\partial S(\eta_i, -I_p)}{\partial \varepsilon} + t_s, \tag{9}$$

where we omit the sub-index + for the phase S_+. The initial coordinate η_i corresponds to the position where the tunneling part of the wave packet is maximum at t_s. The initial η_i value are found solving numerically the time dependent Schrödinger equation in parabolic coordinates. They correspond to the saddle point of the potential.

Transforming back to Cartesian coordinates, we show the Wigner trajectories for the two different regimes, low and high intensity in Fig. 4.

We observe that in both regimes the electron spends a finite time under the barrier until it reaches the exit and it is set into the continuum with a positive time delay $\tau_W \equiv t_W(x_{exit}) > 0$ (see the orange dotted curves in Fig. 4). We also observe a nonvanishing longitudinal momentum along the laser field direction at the tunnel exit $p_W = (dt_W(x)/dx)^{-1}|_{x=x_{exit}}$. This is in agreement with previous observations of the dominant trajectories using a somewhat simpler potential [7, 32].

3.2. Classical propagation

To connect the tunneling step with the subsequent electron motion in the classically allowed region during the second step, we define a classical trajectory starting at the tunnel exit x_{exit} at time $t_{exit} = \tau_W$ with an initial momentum $p_{exit} = p_W$ (Fig. 4 blue curves). As the laser field rotates during the tunneling time delay, the electron emission direction as well as the position of the tunnel exit are adapted to the field direction after the time delay τ_W. The final electron

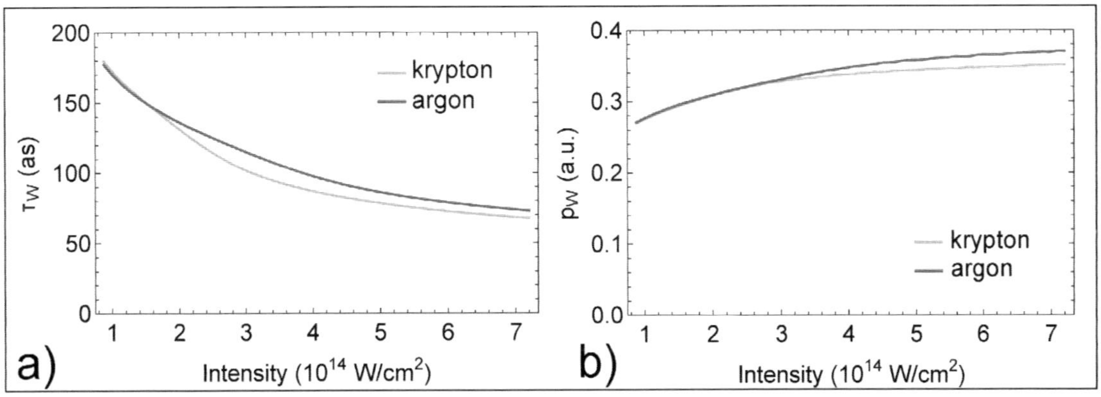

Figure 5. Initial conditions arising from the Wigner formalism for the two investigated atoms (argon and krypton): initial time (a) and initial longitudinal momentum (b). Notably, at the tunnel exit, the differences between both targets, argon and krypton, are as small as 10 attoseconds for the time delay and less than 0.1 a.u. for the longitudinal momentum. Adapted from [23].

momentum after the laser pulse is calculated via classical propagation

$$\boldsymbol{p}_f = \boldsymbol{p}_{exit} - \int_{t_{exit}}^{\infty} dt \; [\boldsymbol{E}(t) + \nabla U(\boldsymbol{r}(t))] \;, \tag{10}$$

by solving Newton's equations of motion $\ddot{\boldsymbol{r}}(t) = -\nabla U(\boldsymbol{r}(t)) - \boldsymbol{E}(\boldsymbol{t})$, with $\boldsymbol{E}(t)$ being the laser electric field and $U(\boldsymbol{r})$ the potential of the atomic core taking into account its polarization by the laser [33].

How the initial conditions affect the asymptotic electron momentum is shown fig. 4 where we compare the trajectory using the Wigner initial condition $t_c(x, \tau_W, p_W)$ (green curve) with the trajectory of the commonly used classical simple-man (SM) model $t_{sm}(x)$. The latter is defined having a zero delay time and zero initial velocity at the tunnel exit. By neglecting the atomic Coulomb potential in the SM model, the final momentum becomes $\boldsymbol{p}_f = -\int_{t_0}^{\infty} dt\, \boldsymbol{E}(t)$ and quantitative information can be derived. A tunneling time delay corresponds to a rotation of the asymptotic momentum distribution by $\delta\theta_\tau = \omega\tau_W$, and the non-zero initial momentum to a counter-rotation by $\delta\theta_p \approx -p_w/p_E$ with $p_E = E_0/\omega$, with ω and E_0 being the laser-field frequency and amplitude, respectively. For low intensities, one compensates the other almost perfectly and the SM trajectory merge asymptotically with the Wigner one. However for larger intensities the Wigner time delay decreases faster than the initial momentum increases. This leads to an additional net rotation compared to the SM prediction $\delta\theta = \delta\theta_\tau - \delta\theta_p$.

4. Comparison between experiment and theory

In Fig. 5 we show the time delay and initial momentum obtained from the Wigner formalism as a function of intensity for each target. These conditions are used to compare with experimental results over a large intensity span.

However, in order to compare experiment and theory, we need to identify for the theory, which pulse parameters should be used (pulse duration, shape and ellipticity) since these parameters are determined in the experiment with a limited accuracy. For these we make use of two additional observables dependent on the laser interaction: the difference in average radius $\rho_{Ar} - \rho_{Kr}$ and the yield ratio $Yield_{Ar} - Yield_{Kr}$ and analyze them as a function of intensity (Fig. 6). We

IOP Conf. Series: Journal of Physics: Conf. Series **999** (2018) 012004 doi:10.1088/1742-6596/999/1/012004

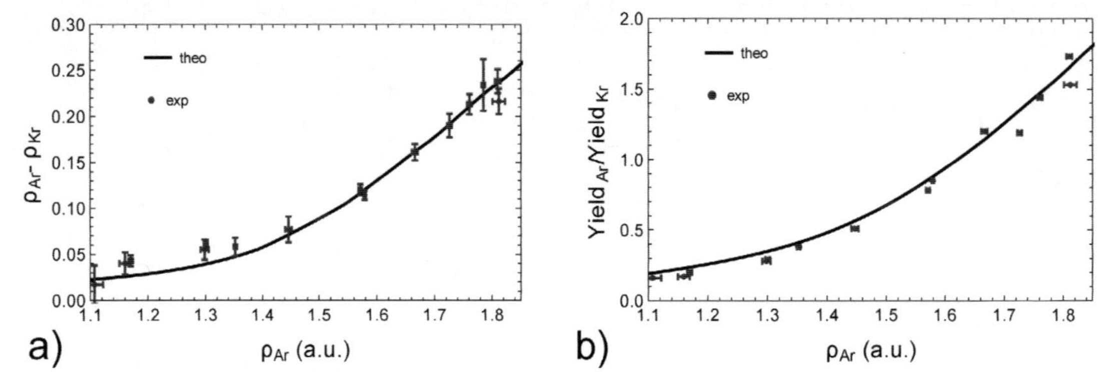

Figure 6. Comparison of the experiment and the theory; a) the difference between the average electron momentum, b) the yield ratio. Adapted from [23].

extract the pulse parameters which give the best agreement between theory and experiment for these two observables.

With these parameters for the theory we compare the angle difference between argon and krypton as shown in Fig. 7. We note that the only differences between the two theoretical curves in Fig. 7 are the chosen combinations for the initial momentum and the Wigner time. We conclude that the simple-man model (the dashed curve) fails to predict the experimental data whereas the solid curve, whose initial conditions are given by the Wigner formalism, agrees well with the experiment in the regime where $\rho_{Ar} \geq 1.3a.u.$ [23]. In the low intensity domain ($\rho_{Ar} < 1.3a.u.$) one has to take into account non-adiabatic effects to tunnel-ionization such as non-adiabatic momentum shifts [34], and tunnel exit shifts [18]. This is the domain where the Wigner formalism converges to the simple-man model because the Keldysh parameter approaches unity.

The agreement remains even if slightly different parameters are used (pulse shape, pulse duration, ellipticity). There is no scenario with the initial time and momentum set to zero which can reproduce the observed angular distribution. The variation of pulse parameters has almost no influence on the main observable $\Delta(\theta_{max})$.

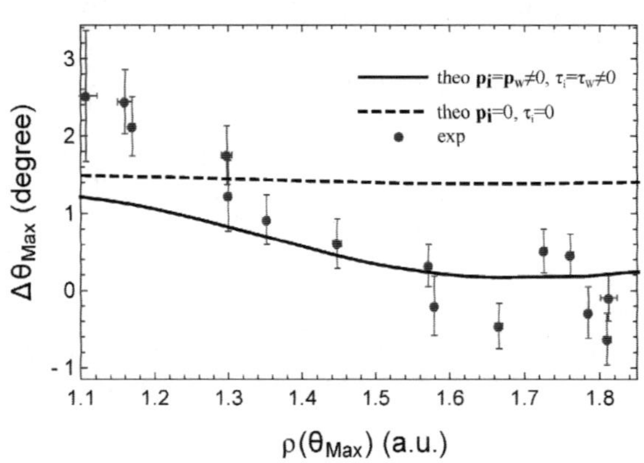

Figure 7. Difference between the most probable photoelectron emission angle for argon and krypton. Experiment and theory (with and without initial momentum and tunneling delay time). Adapted from [23].

IOP Conf. Series: Journal of Physics: Conf. Series **999** (2018) 012004 doi:10.1088/1742-6596/999/1/012004

5. Conclusion

We have investigated the quasi-classical description of strong-field tunneling ionization. Based on the Wigner formalism we identify the tunneling delay time and the initial longitudinal momentum at the tunnel exit for the most probable trajectory of the tunneling electron and test this formalism by comparison to the experiment. We show that the experimental results can be reproduced only when nonvanishing values for these parameters are used.

References

[1] Brabec T and Krausz F 2000 *Rev. Mod. Phys.* **72** 545–591
[2] Keldysh L V 1965 *JETP* **47** 1307
[3] Eckle P, Pfeiffer A N, Cirelli C, Staudte A, Dörner R, Muller H G, Büttiker M and Keller U 2008 *Science* **322** 1525–1529
[4] Krausz F and Ivanov M 2009 *Rev. Mod. Phys.* **81**(1) 163–234
[5] Pfeiffer A N, Cirelli C, Landsman A S, Smolarski M, Dimitrovski D, Madsen L B and Keller U 2012 *Phys. Rev. Lett.* **109**(8) 083002
[6] Landsman A S, Weger M, Maurer J, Boge R, Ludwig A, Heuser S, Cirelli C, Gallmann L and Keller U 2014 *Optica* **1** 343–349
[7] Yakaboylu E, Klaiber M and Hatsagortsyan K Z 2014 *Phys. Rev. A* **90**(1) 012116
[8] Ni H, Saalmann U and Rost J M 2016 *Phys. Rev. Lett.* **117**(2) 023002
[9] Torlina L, Morales F, Kaushal J, Ivanov I, Kheifets A, Zielinski A, Scrinzi A, Muller H G, Sukiasyan S, Ivanov M and Smirnova O 2015 *Nature Phys.* **11** 503–508
[10] Corkum P B and Krausz F 2007 *Nature Phys* **3** 381
[11] Calegari F, Sansone G, Stagira S, Vozzi C and Nisoli M 2016 *J. Phys. B* **49** 062001
[12] Meckel M, Comtois D, Zeidler D, Staudte A, Pavičić D, Bandulet H C, Pépin H, Kieffer J C, Dörner R, Villeneuve D M and Corkum P B 2008 *Science* **320** 1478–1482
[13] Blaga C I, Xu J, DiChiara A D, Sistrunk E, Zhang K, Agostini P, Miller T A, DiMauro L F and Lin C D 2012 *Nature* **483** 194–197
[14] Eckle P, Smolarski M, Schlup P, Biegert J, Staudte A, Schöffler M, Muller H G, Dörner R and Keller U 2008 *Nature Phys.* **4** 565–570
[15] Pfeiffer A N, Cirelli C, Smolarski M, Dimitrovski D, Abu-samha M, Madsen L B and Keller U 2012 *Nature Phys* **8** 76–80
[16] Barth I and Smirnova O 2011 *Phys. Rev. A* **84**(6) 063415
[17] Li M, Liu Y, Liu H, Ning Q, Fu L, Liu J, Deng Y, Wu C, Peng L Y and Gong Q 2013 *Phys. Rev. Lett.* **111**(2) 023006
[18] Klaiber M, Hatsagortsyan K Z and Keitel C H 2015 *Phys. Rev. Lett.* **114**(8) 083001
[19] Li M, Geng J W, Han M, Liu M M, Peng L Y, Gong Q and Liu Y 2016 *Phys. Rev. A* **93**(1) 013402
[20] Wigner E P 1955 *Phys. Rev.* **98**(1) 145–147
[21] Ullrich J, Moshammer R, Dorn A, Dörner R, Schmidt L P H and Schmidt-Böcking H 2003 *Reports on Progress in Physics* **66** 1463
[22] De Jesus V, Rudenko A, Feuerstein B, Zrost K, Schröter C, Moshammer R and Ullrich J 2004 *Journal of electron spectroscopy and related phenomena* **141** 127–142
[23] Camus N, Yakaboylu E, Fechner L, Klaiber M, Laux M, Mi Y, Hatsagortsyan K Z, Pfeifer T, Keitel C H and Moshammer R 2017 *Phys. Rev. Lett.* **119**(2) 023201
[24] Alnaser A S, Tong X M, Osipov T, Voss S, Maharjan C M, Shan B, Chang Z and Cocke C L 2004 *Phys. Rev. A* **70**(2) 023413
[25] Ivanov I A and Kheifets A S 2014 *Phys. Rev. A* **89**(2) 021402
[26] Boge R, Cirelli C, Landsman A S, Heuser S, Ludwig A, Maurer J, Weger M, Gallmann L and Keller U 2013 *Phys. Rev. Lett.* **111**(10) 103003
[27] Schulman L S 2012 *Techniques and applications of path integration* (Dover Publications)
[28] Teeny N, Yakaboylu E, Bauke H and Keitel C H 2016 *Phys. Rev. Lett.* **116**(6) 063003
[29] Byron F W and Fuller R W 1992 *Mathematics of classical and quantum physics* (New York,: Dover Publications Inc.)
[30] Arfken G B 2013 *Mathematical methods for physicists* (New York: Academic press)
[31] Morse P and Feshbach H 1953 *Methods of theoretical physics* International series in pure and applied physics (McGraw-Hill)
[32] Czirjk A, Kopold R, Becker W, Kleber M and Schleich W 2000 *Optics Communications* **179** 29 – 38
[33] Dimitrovski D, Martiny C P J and Madsen L B 2010 *Phys. Rev. A* **82**(5) 053404
[34] Mur V, Popruzhenko S and Popov V 2001 *Journal of Experimental and Theoretical Physics* **92** 777–788

IOP Conf. Series: Journal of Physics: Conf. Series **999** (2018) 012005 doi:10.1088/1742-6596/999/1/012005

Interference-Free and Interference-Dominated Photoionization: Synthesis of Ultrashort and Coherent Single-Electron Wave Packets

F Cajiao Vélez, J Z Kamiński and K Krajewska

Institute of Theoretical Physics, Faculty of Physics, University of Warsaw, Pasteura 5, 02-093 Warsaw, Poland

E-mail: Felipe.Cajiao-Velez@fuw.edu.pl

Abstract. Ionization of hydrogen-like ions driven by intense, short, and circularly-polarized laser pulses is considered under the scope of the relativistic strong-field approximation. We show that the energy spectra of photoelectrons can exhibit two types of structures, i.e., interference-dominated or interference-free ones. These structures are analyzed in connection to the time-dependent ponderomotive energy of electrons in the laser field. A possibility of synthesis of ultrashort single-electron pulses from those structures is also investigated.

1. Introduction

In four-dimensional electron microscopy and diffraction (see, for instance, Refs. [1, 2, 3, 4] and the reviews [5, 6, 7, 8]), short electron wave packets are used to monitor dynamical changes occurring on the femtosecond and picometer scales. Furthermore, as pulses comprising of many electrons tend to spread fast in time, which is due to space-charge effects, a lot of effort has been devoted to the generation of single-electron pulse sources [3, 4, 5, 7, 9, 10, 11].

Recently, we have demonstrated that ultrashort and coherent single-electron wave packets can be obtained during ionization driven by circularly-polarized and intense laser pulses [12, 13, 14, 15]. We have shown that the shape of the temporal ponderomotive energy curve allows one to predict the polar and azimuthal angles at which the electron pulses can be synthesized. Also, the time-duration of the resulting electron wave packets can be significantly modified by changing the intensity of the laser pulse or its polarization [16, 17].

It is the aim of this paper to further analyze the relation between the probability distribution of photoionization and the time-dependent ponderomotive energy. This is done in Sec. 3, after the necessary theoretical formulation is given in Sec. 2. Furthermore, in Sec. 4, we demonstrate how different patterns in the energy spectra of photoelectrons may lead to the synthesis of very short and coherent single-electron wave packets. Finally, in Sec. 5, we present our conclusions and remarks.

In the following, we use units such that $\hbar = 1$. The Feynman slash notation $\not{a} = \gamma \cdot a = \gamma^\mu a_\mu$, where a is an arbitrary four-vector and γ are the Dirac gamma matrices, is also employed. In Sec. 2 we make use of the so-called light-cone variables, which are defined with respect to a unit vector $\boldsymbol{n}$. For instance, for the space-time four-vector $x = (ct, \boldsymbol{x}) \equiv (x^0, \boldsymbol{x})$, such variables are:

IOP Conf. Series: Journal of Physics: Conf. Series **999** (2018) 012005 doi:10.1088/1742-6596/999/1/012005

$x^{\parallel} = \boldsymbol{n} \cdot \boldsymbol{x}$, $x^{-} = x^0 - x^{\parallel}$, $x^{+} = (x^0 + x^{\parallel})/2$, and $\boldsymbol{x}^{\perp} = \boldsymbol{x} - x^{\parallel}\boldsymbol{n}$. Furthermore, the average of a function $F(\phi)$ over the laser pulse duration is denoted with the angle brackets, i.e., $\langle F \rangle$.

2. Theory

Consider a hydrogen-like ion interacting with an intense, short, and circularly-polarized laser pulse represented by the four-vector potential $A(x)$. The latter, in the so-called plane-wave-fronted pulse approximation, can be written in terms of two arbitrary shape functions, $f_1(\phi)$ and $f_2(\phi)$, which depend on the laser phase, $\phi = k \cdot x$, and vanish for $\phi < 0$ and $\phi > 2\pi$. Namely,

$$A(k \cdot x) = A_0[\varepsilon_1 f_1(k \cdot x) + \varepsilon_2 f_2(k \cdot x)], \tag{1}$$

where A_0 is the amplitude of the vector potential and $k = k^0(1, \boldsymbol{n}) \equiv k^0 n$ is the wave four-vector. Here, $\boldsymbol{n}$ is a unit vector which represents the direction of propagation of the laser field, $k^0 = \omega/c$ is the wave-number, $\omega = 2\pi/T_p$ is the fundamental frequency of laser field oscillations, and T_p is the pulse duration. Also, in (1), $\varepsilon_i = (0, \boldsymbol{\varepsilon}_i)$ $i = 1, 2$ are two polarization four-vectors such that $\varepsilon_i \cdot \varepsilon_j = -\delta_{ij}$ and $k \cdot \varepsilon_i = 0$. Due to the interaction with the laser field, the ion being initially in the bound state $\Psi_{\rm i}(\boldsymbol{x})$ of energy E_0 and the spin degree of freedom $\lambda_{\rm i} = \pm$, gets ionized. The probability amplitude of ionization, in the framework of the relativistic strong-field approximation (RSFA) [12, 18, 19, 20], which is based *solely* on the lowest-order Born expansion of the final scattering state with respect to the bounding potential, is given by

$$\mathcal{A}_{\lambda_{\rm i}\lambda}(\boldsymbol{p}) = -{\rm i} \int {\rm d}^4 x \, {\rm e}^{-{\rm i}(E_0/c)x^0} \bar{\psi}^{(+)}_{\boldsymbol{p}\lambda}(x) e \not{A}(x) \Psi_{\rm i}(\boldsymbol{x}), \tag{2}$$

where $\psi^{(+)}_{\boldsymbol{p}\lambda}(x)$ is the Volkov state [21, 22, 23]. Such state describes the quantum-mechanical behavior of a free electron of asymptotic momentum $\boldsymbol{p}$, energy $E_{\boldsymbol{p}} = \sqrt{(m_{\rm e}c^2)^2 + (c\boldsymbol{p})^2}$, and the spin degree of freedom $\lambda = \pm$ in the laser pulse (1).

The total probability of photoionization, with fixed λ and $\lambda_{\rm i}$, can be calculated by integrating the modulus squared of the amplitude (2) over the density of final electron states. This allows us to define the spin resolved triply-differential probability distribution which, in atomic units, is represented as $\mathcal{P}_{\lambda_{\rm i}\lambda}(\boldsymbol{p})$. Also, by summing the values corresponding to the final spin states while averaging over the initial ones, we obtain the spin-independent probability distribution $\mathcal{P}(\boldsymbol{p})$, i.e.,

$$\mathcal{P}(\boldsymbol{p}) = \frac{1}{2} \sum_{\lambda_{\rm i}, \lambda = \pm} \mathcal{P}_{\lambda_{\rm i}\lambda}(\boldsymbol{p}). \tag{3}$$

Even though we do not present here the explicit formula defining this quantity, it can be found in Eqs. (18) and (21) of Ref. [14].

It is worth mentioning that, if the lowest-order Born expansion is used, the application of the RSFA is justified provided that the photoelectron kinetic energy is much larger than the ionization potential of the field-free ionic system, i.e.,

$$\sqrt{(m_{\rm e}c^2)^2 + (c\boldsymbol{p})^2} - m_{\rm e}c^2 \gg m_{\rm e}c^2 - E_0. \tag{4}$$

Thus, the analysis of the low-energy portion of the probability distributions using the RSFA could be only qualitative. For this reason, in this paper we shall concentrate on the high-energy ionization, in which Eq. (4) holds.

For our further considerations, we introduce here the time-dependent ponderomotive energy of the electron in the laser field. In terms of the vector potential (1), it is given by

$$U_p(\phi) = \frac{e^2 \boldsymbol{A}^2(\phi)}{2m_{\rm e}} = \frac{e^2 A_0^2}{2m_{\rm e}} \left[f_1^2(\phi) + f_2^2(\phi) \right]. \tag{5}$$

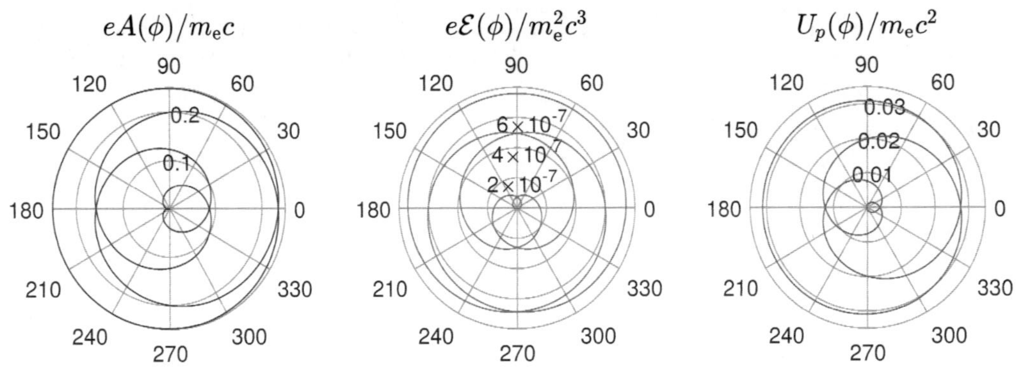

Figure 1. Evolution of the tips of the vector potential $\boldsymbol{A}(\phi)$ (left panel) and electric field $\boldsymbol{\mathcal{E}}(\phi)$ (middle panel) for the laser pulse defined by (6). The corresponding time-dependent ponderomotive energy [Eq. (5)] is also shown (right panel). The laser field parameters are $I = 10^{17}\,\mathrm{W/cm}^2$, $N_{\rm osc} = 5$, and $\omega_{\rm L} = N_{\rm osc}\omega = 1.5498\,\mathrm{eV}$ (see the text). All figures are presented in relativistic units in the xy-plane, when changing the laser field phase ϕ from 0 to 2π.

As it will become clear shortly, this expression will be helpful in understanding the general features of the energy spectra of photoelectrons (see, Sec. 3.1).

2.1. Laser field

For our numerical illustrations, we consider a laser pulse consisting of $N_{\rm osc}$ field cycles within a $\sin^2$ envelope interacting with a hydrogen-like ion of atomic number $Z = 2$ (He^+). The electric field characterizing such pulse, in the plane-wave-fronted pulse approximation, is determined by the shape functions

$$F_j(\phi) = N_0 \sin^2\Big(\frac{\phi}{2}\Big)\sin(N_{\rm osc}\phi + \delta_j + \chi)\cos(\delta + \delta_j), \quad \text{for } j = 1, 2. \tag{6}$$

As mentioned before, these functions are defined in the interval $0 \leqslant \phi \leqslant 2\pi$ and are zero otherwise. Here, δ and δ_j define the polarization of the laser field, χ is the carrier-envelope phase, and $N_0 = \sqrt{8/3}N_{\rm osc}$ is a normalization constant [12]. In the following, we assume a circular polarization [$\delta = \pi/4$, $\delta_j = (j-1)\pi/2$] for a laser pulse propagating along the z-direction ($\boldsymbol{\varepsilon}_1 = \boldsymbol{e}_x$, $\boldsymbol{\varepsilon}_2 = \boldsymbol{e}_y$ and $\boldsymbol{n} = \boldsymbol{e}_z$,). Also, the carrier-envelope phase is $\chi = \pi$.

As the electric field and the vector potential obey the relation $\boldsymbol{\mathcal{E}}(\phi) = -\partial_t \boldsymbol{A}(\phi)$, the shape functions defining the latter [see, Eq. (1)] can be obtained from (6). Namely,

$$f_j(\phi) = -\int_0^{\phi} \mathrm{d}\phi'\, F_j(\phi'), \tag{7}$$

and they also vanish outside the interval $0 \leqslant \phi \leqslant 2\pi$.

In Fig. 1, we show the parametric plots of the tips of the vector potential (left panel) and electric field (middle panel) describing the laser pulse (6). The parameter used to obtain those curves, which is the phase ϕ, changes from 0 to 2π. Also, in the right panel of the same figure, the time-dependent ponderomotive energy $U_p(\phi)$ [Eq. (5)] is plotted. Note that, in Fig. 1 the carrier frequency corresponds to $\omega_{\rm L} = N_{\rm osc}\omega = 1.5498\,\mathrm{eV}$, the number of field oscillations is $N_{\rm osc} = 5$, and the time-averaged intensity is $I = 10^{17}\,\mathrm{W/cm}^2$. All curves are presented in relativistic units, they start at the origin of coordinates and evolve counterclockwise while increasing ϕ.

In the following, we define the azimuthal angle φ in terms of the vector potential. Namely,

$$\varphi \equiv \varphi(\phi) = \arg\big(e\boldsymbol{A}(\phi)\cdot\boldsymbol{\varepsilon}_1 + \mathrm{i}e\boldsymbol{A}(\phi)\cdot\boldsymbol{\varepsilon}_2\big), \tag{8}$$

where $0 \leqslant \arg(z) < 2\pi$ represents the argument of the complex number z. According to this definition, φ is not a one-to-one function of the phase ϕ. For this reason, all curves presented in Fig. 1 intersect themselves at different phases. That will result in interference patterns observed in the ionization distributions (see, the next Section).

3. Probability distribution in strong-field photoionization

To calculate the probability distribution of ionization directly from Eq. (2) requires a considerable computational effort. For this reason, in recent publications (see, e.g., Refs. [14, 15]), the saddle-point approximation has been applied in (2), which greatly simplifies and speeds up numerical calculations. With this method, certain properties of the probability distributions, when calculated in a given space direction, were determined. The latter is defined by the polar and azimuthal angles of electron detection $\theta_{\boldsymbol{p}}$ and $\varphi_{\boldsymbol{p}}$, respectively. For the purpose of this paper, we shall only refer to the aforementioned properties and their consequences. Their more detailed description can be found in Refs. [12, 14, 15, 16, 17].

First, it has been shown that the energy spectra of photoelectrons can exhibit two types of structures; namely, interference-dominated and interference-free ones [12, 14, 15, 17]. While the former are characterized by dense oscillations, the latter appear as broad lobes spanning over hundreds of single-photon energies. Furthermore, when the driving pulse is linearly polarized or lasts for a very long time, only interference-dominated structures are expected [17]. In contrast, when the pulse is short and its polarization is circular or, in general, elliptical, interference-free structures appear [12, 14, 15, 16, 17]. The interference-free structures are also known as supercontinua in photoionization.

Second, if the laser pulse is not linearly-polarized, the target ion is sufficiently light ($Z\alpha \ll 1$), and the photoelectrons are fast enough ($E_{\boldsymbol{p}} - m_{\mathrm{e}}c^2 \gg m_{\mathrm{e}}c^2 - E_0$), there exist a real phase $\phi_{\boldsymbol{p}}$ which determines several properties of the probability distribution of photoelectrons. Such phase, which typically is associated with the time when the electron appears in the continuum, relates to the azimuthal angle of electron detection $\varphi_{\boldsymbol{p}}$ according to [16, 17]

$$\varphi_{\boldsymbol{p}} = \arg\big(e\boldsymbol{A}(\phi_{\boldsymbol{p}}) \cdot \boldsymbol{\varepsilon}_1 + \mathrm{i}e\boldsymbol{A}(\phi_{\boldsymbol{p}}) \cdot \boldsymbol{\varepsilon}_2\big). \tag{9}$$

As the vector potential $\boldsymbol{A}(\phi)$ is not a uniquely-defined function of the azimuthal angle φ [cf. Eq. (8)], it is possible to find different phases $\phi_{\boldsymbol{p}}$ which, for a given $\varphi_{\boldsymbol{p}}$, satisfy Eq. (9).

Third, it has been shown that there is a close relation between the central energy of the supercontinuum and the ponderomotive energy of the electron in the laser field at its time of birth. Namely, the supercontinua are approximately located at the kinetic energies given by the relation [16, 17]

$$E_{\boldsymbol{p}} - m_{\mathrm{e}}c^2 \approx U_p(\phi_{\boldsymbol{p}}). \tag{10}$$

However, as the $U_p(\varphi)$-spiral may intersect itself at several points (see, the right panel of Fig. 1), two or more phases $\phi_{\boldsymbol{p}}$ can contribute to the probability of electron being detected in the given direction and with the given kinetic energy. Hence, the self-intersections of the ponderomotive energy curve lead to the formation of interference-dominated structures in the energy spectra of photoelectrons. In contrast, for azimuthal angles for which the arcs of the spiral do not intersect, supercontinua will be observed.

Finally, it is worth mentioning that, for the geometry considered in this paper, the RSFA predicts a shift of the direction of photoelectron detection towards polar angles smaller than $\pi/2$. This shift depends on the final kinetic energy of the electrons, as the most energetic ones absorb a larger number of laser photons and, therefore, experience stronger radiation pressure [24]. Moreover, according to [16], it can be estimated that the maximum value of the probability

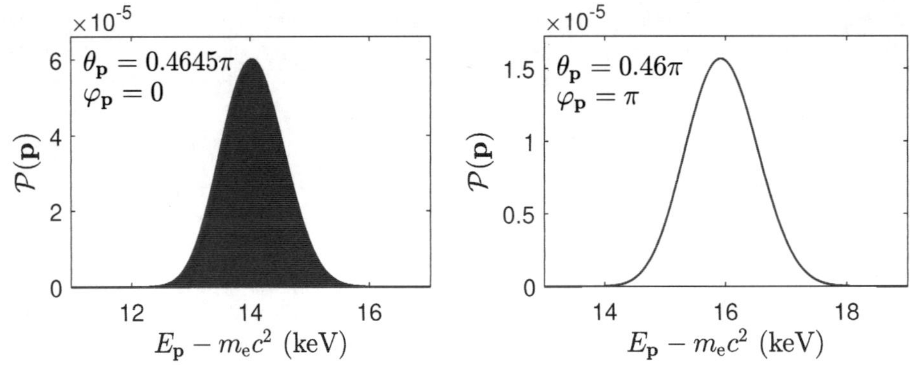

Figure 2. Spin-independent energy spectra of photoelectrons $\mathcal{P}(\boldsymbol{p})$ calculated for the same laser field parameters as in Fig. 1. The polar and azimuthal angles of detection are specified in each panel.

distribution is observed at the polar angle,

$$\theta_{\boldsymbol{p}} \approx \arccos\sqrt{\frac{E_{\boldsymbol{p}} - m_e c^2}{E_{\boldsymbol{p}} + m_e c^2}}. \tag{11}$$

As before, such expression is valid provided that the target ions are light enough and the photoelectrons are fast.

Note that, by setting the azimuthal angle of detection and determining the phase (or phases) $\phi_{\boldsymbol{p}}$ which satisfy (9), several features of the probability distribution can be predicted. For instance, Eq. (10) determines the most probable final kinetic energy of photoelectrons. Also, it can be inferred from the $U_p(\varphi)$-spiral whether or not interference effects would take place. Finally, Eq. (11) determines the direction in space where photoelectrons with a given energy will be most likely observed. In the remaining part of this Section, we shall make use of those observations to analyze the probability distributions.

3.1. Energy spectra of photoelectrons

We now illustrate our theory by calculating the spin-independent probability distribution in photoionization. Fig. 2 shows the energy spectra of photoelectrons $\mathcal{P}(\boldsymbol{p})$ calculated at a given direction in space, as defined by the pair of angles $(\theta_{\boldsymbol{p}}, \varphi_{\boldsymbol{p}})$, and for the laser pulse described in Fig. 1. While the azimuthal angle is chosen arbitrarily, the polar angle is set such that the distribution acquires its maximum values. From Fig. 2 one can see that, for $(\theta_{\boldsymbol{p}}, \varphi_{\boldsymbol{p}}) = (0.4645\pi, 0)$ (left panel), an interference-dominated structure within a Gaussian-like envelope appears. Its maximum is located at 14 keV whereas its energy width $(\Delta E_{\boldsymbol{p}})$, calculated as the full-width at half maximum (FWHM), is roughly 1.25 keV. For this specific structure, the ratio $\Delta E_{\boldsymbol{p}}/E^{\text{center}}$ (also known as the degree of monochromaticity [3, 11]), where E^{center} represents its central kinetic energy, is approximately 0.09. Now, in the right panel of Fig. 2, where $(\theta_{\boldsymbol{p}}, \varphi_{\boldsymbol{p}}) = (0.46\pi, \pi)$, a broad supercontinuum is observed. In this case, the maximum of the distribution is located at 15.9 keV, its width is $\Delta E_{\boldsymbol{p}} \approx 1.4$ keV, and the ratio $\Delta E_{\boldsymbol{p}}/E^{\text{center}}$ is also approximately 0.09.

Let us now analyze the relation between the position of maxima in the high-energy portion of the photoelectron spectra and the shape of the ponderomotive energy curve (right panel of Fig. 1). The latter intersects itself at $\varphi_{\boldsymbol{p}} = \varphi = 0$, which corresponds to the energy $0.0276 m_e c^2 \approx 14.2$ keV. Hence, according to the observations mentioned above, at $\varphi_{\boldsymbol{p}} = 0$,

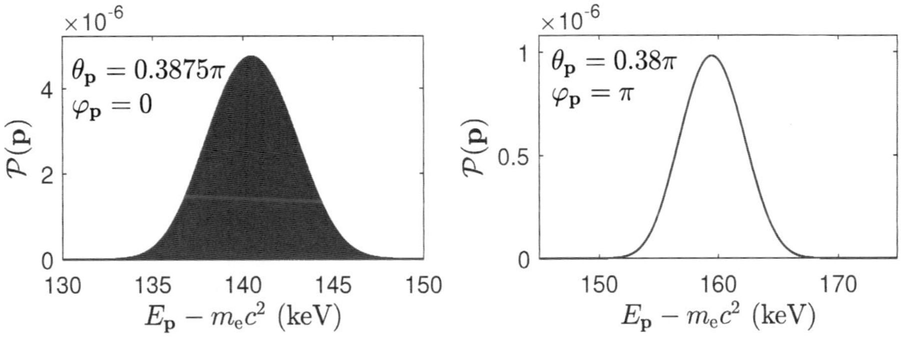

Figure 3. The same as in Fig. 2 but for the time-averaged laser field intensity $I = 10^{18}\,\mathrm{W/cm^2}$. The polar and azimuthal angles of detection are shown in the upper part of both panels.

an interference-dominated structure with maxima at around 14 keV is expected. This is exactly the case, as shown in the left panel of Fig. 2. In contrast, the line $\varphi_p = \varphi = \pi$, crosses a single external arc of the $U_p(\varphi)$-curve at $0.0312 m_e c^2 \approx 16$ keV. This explains the formation of the supercontinuum observed in the right panel of Fig. 2 at roughly 16 keV.

Note that the predictions arising from Eq. (10) go even further. For instance, if we increase the laser field intensity by one order of magnitude, the maximum of the probability distribution of ionization will appear at ten times larger kinetic energies [see, also Eq. (5)]. As the ponderomotive energy curve scales linearly with the driving field intensity while conserving its general shape, it is also expected that the interference-free and interference-dominated structures will be located at the same azimuthal angles $\varphi_{\boldsymbol{p}}$. In order to check that, in Fig. 3 we present the spin-independent energy spectra of photoelectrons $\mathcal{P}(\boldsymbol{p})$ calculated at the same azimuthal angles as in Fig. 2 but for the laser field intensity $I = 10^{18}\,\mathrm{W/cm^2}$. As expected, while at $\varphi_{\boldsymbol{p}} = 0$ (left panel) we observe an interference-dominated structure with maximum at 140 keV, at $\varphi_{\boldsymbol{p}} = \pi$ (right panel) the probability distribution exhibits a broad supercontinuum with maximum at 159 keV. This is in perfect agreement with the predictions arising from Eq. (10). Now, with respect to the bandwidth of each structure, it can be estimated that the interference-dominated pattern has the energy spread of roughly 5.8 keV and the ratio $\Delta E_{\boldsymbol{p}}/E^{\mathrm{center}} \approx 0.04$. On the other hand, while the supercontinuum exhibits a larger bandwidth, $\Delta E_{\boldsymbol{p}} \approx 6.4$ keV, the ratio $\Delta E_{\boldsymbol{p}}/E^{\mathrm{center}}$ is also around 0.04. By comparing Figs. 2 and 3, one can conclude that the width of the high-probability structures increase with the laser pulse intensity. However, it does not change with intensity at the same rate as the central kinetic energy does. Moreover, one can expect that electron pulses with smaller degree of monochromaticity can be obtained in high-energy ionization by further increasing the intensity of the driving field. This is due to the fact that the ratio $\Delta E_{\boldsymbol{p}}/E^{\mathrm{center}}$ continuously decreases with this parameter.

In closing this Section, let us note that the polar angles at which we observe the maximum probability of ionization decrease from $\theta_{\boldsymbol{p}} \approx 0.46\pi$, for the intensity $I = 10^{17}\,\mathrm{W/cm^2}$, to $\theta_{\boldsymbol{p}} \approx 0.38\pi$, for $I = 10^{18}\,\mathrm{W/cm^2}$ (see, Figs. 2 and 3). Such values agree amazingly well with the predictions arising from Eq. (11).

4. Synthesis of ultrashort and coherent single-electron wave packets

In this Section, we will show how the interference-dominated or interference-free structures can be used to synthesize very short and coherent single-electron pulses. Note that, according to the uncertainty principle, the short-in-time electron wave packets are characterized by large energy bandwidths. Therefore, for our current purpose, we shall concentrate on the structures located in the high-energy portion of the probability distributions, as they are characterized by the

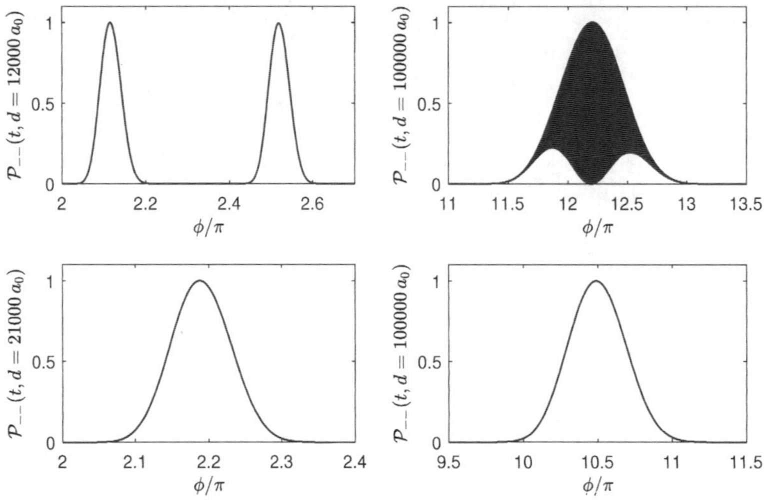

Figure 4. Space-time profiles $\mathcal{P}_{--}(t,d)$ calculated from Eq. (13) for the pairs of angles $(\theta_0, \varphi_0) = (0.4645\pi, 0)$ (upper row) and $(0.46\pi, \pi)$ (lower row). Such distributions are obtained at a distance d from the parent ion and correspond to the energy spectra of photoelectrons shown in the left and right panels of Fig. 2, respectively. The kinetic energy boundaries are $(E_{\rm min}, E_{\rm max})$=(11 keV, 17 keV) for the upper panels and (13 keV,19 keV) for the lower ones.

largest energy spread.

The spin-dependent space-time probability amplitude $\mathcal{A}_{\lambda_i\lambda}(t,d)$, which defines the probability of finding a single-electron wave packet propagating along the direction $\boldsymbol{n}_0$ and at a distance $d = \boldsymbol{x}\cdot\boldsymbol{n}_0$ from the parent ion, is given by [12, 14, 15, 16, 17]

$$\mathcal{A}_{\lambda_i\lambda}(t,d) = N_{\mathcal{A}} e^{i m_e c^2 t} \int_{E_{\rm min}+m_e c^2}^{E_{\rm max}+m_e c^2} dE_{\boldsymbol{p}} e^{-i(E_{\boldsymbol{p}}t - |\boldsymbol{p}|d)} u_{|\boldsymbol{p}|\boldsymbol{n}_0,\lambda} \sqrt{E_{\boldsymbol{p}}} |\boldsymbol{p}| \mathcal{A}_{\lambda_i\lambda}(|\boldsymbol{p}|\boldsymbol{n}_0), \qquad (12)$$

where $\mathcal{A}_{\lambda_i\lambda}(\boldsymbol{p})$ is the probability amplitude of photoionization (2), $N_{\mathcal{A}}$ is a normalization constant, and $E_{\rm min}$ and $E_{\rm max}$ select the kinetic energy range from which the wave packet is synthesized. Moreover, the spin-dependent space-time probability distribution is calculated from (12) as

$$\mathcal{P}_{\lambda_i\lambda}(t,d) = [\mathcal{A}_{\lambda_i\lambda}(t,d)]^\dagger \mathcal{A}_{\lambda_i\lambda}(t,d). \qquad (13)$$

Here, the initial and final spin states are projected onto the laser field propagation direction, $\boldsymbol{e}_z$.

In the upper and lower rows of Fig. 4, we present the temporal profiles $\mathcal{P}_{--}(t,d)$ [Eq. (13)] obtained from the energy spectra of photoelectrons shown in the left and right panels of Fig. 2, respectively. Here, the directions of propagation, defined by the pair of angles (θ_0, φ_0), are $(0.4645\pi, 0)$ (upper row) and $(0.46\pi, \pi)$ (lower row). While in the left column the distances from the parent ion are such that the photoelectron just left the laser pulse (i.e., $\phi = k\cdot x \gtrsim 2\pi$), the right column relates to a given distance of $1\times10^5 a_0 \approx 5.3\ \mu$m, where a_0 is the Bohr radius. Note that, as the *normalized* space-time distributions behave in the same way independently of the initial and final spin states, only the case $(\lambda_i, \lambda) = (-,-)$ is considered.

From the upper-left panel of Fig. 4 it can be seen that, at $d = 1.2\times10^4 a_0 \approx 0.6\ \mu$m, the interference-dominated structure leads to a double-hump temporal profile, characterized by two well-defined Gaussian-like lobes, each one of duration around 440 as. However, at larger distances from the parent ion, both lobes undergo a considerable broadening (upper-right panel). This leads to the formation of a single structure with a rich interference pattern, which is a signature

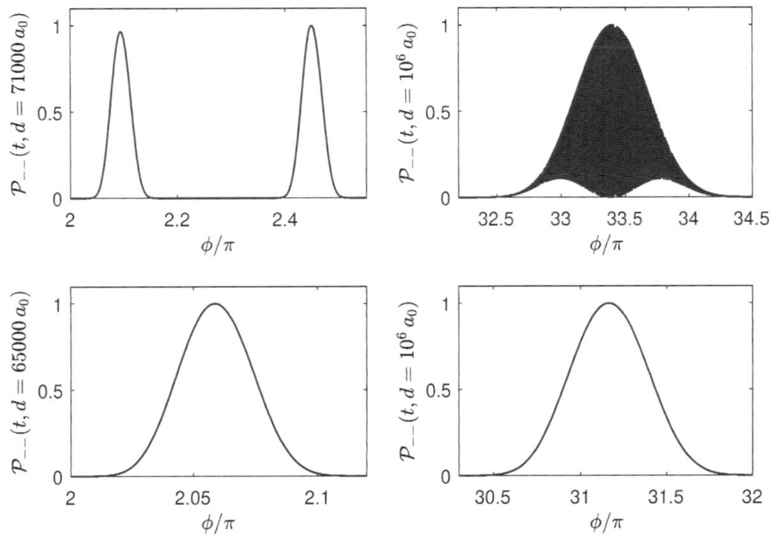

Figure 5. The same as in Fig. 4 but calculated for the energy spectra of photoelectrons shown in Fig. 3 (i.e., for the time-averaged laser intensity $I = 10^{18}\mathrm{W/cm^2}$). The directions of propagation are $(\theta_0, \varphi_0) = (0.3875\pi, 0)$ (upper row) and $(0.38\pi, \pi)$ (lower row). Here, the kinetic energy boundaries are $(E_{\min}, E_{\max})$ =(130 keV,150 keV) for the upper row and (145 keV,175 keV) for the lower one. In the right column we set $d = 10^6 a_0 \approx 53$ μm.

of the coherence of the process. From the FWHM of its envelope, it can be estimated that at $d \approx 5.3$ μm such structure lasts for 4.4 fs. On the other hand, if the electron wave packets are obtained from the supercontinuum shown in the right panel of Fig. 2, we observe a single and smoothly-varying temporal profile. While at a distance $d = 2.1 \times 10^4 a_0 \approx 1.1$ μm from the parent ion, the duration of the electron pulse is approximately 670 as (lower-left panel of Fig. 4), at $d \approx 5.3$ μm it increases up to 3.2 fs (lower-right panel of the same figure). However, even at the larger distance, the electron wave packet still exhibits a smooth, interference-free behavior.

In Fig. 5 we present the same as in Fig. 4 but for the laser field intensity $I = 10^{18}\mathrm{W/cm^2}$. This time, we set $(\theta_0, \varphi_0) = (0.3875\pi, 0)$ (upper row) and $(0.38\pi, \pi)$ (lower row). Note that those temporal profiles relate to the energy spectra of photoelectrons shown in the left and right panels of Fig. 3, respectively. From the upper-left panel of Fig. 5 one can see that, at the distance $d = 7.1 \times 10^4 a_0 \approx 3.8$ μm from the parent ion, a double-hump structure is also formed. As calculated from the FWHM of the Gaussian-like lobes, such structure would lead to an electron pulse lasting only for 307 as (each individual hump). However, at $d \approx 53$ μm a broad interference pattern, with duration of 4.6 fs, is formed (upper-right panel). In contrast, while an electron pulse synthesized from the supercontinuum lasts for 245 as at a distance of 3.4 μm (lower-left panel), it increases its duration up to 3.6 fs at $d = 53$ μm (lower-right panel).

Let us note that the analysis presented in this paper relates to photoelectrons with sub-relativistic kinetic energies, i.e., $E_{\boldsymbol{p}} - m_{\mathrm{e}}c^2 < m_{\mathrm{e}}c^2$. Therefore, it is expected that the synthesized electron pulses spread fast in time. However, by increasing the intensity of the laser field, it is possible to create nearly-relativistic (or relativistic) wave packets of short duration [15]. In such case, the temporal spreading is negligible or very small. Moreover, it was recently demonstrated that it is also possible to obtain very energetic electron pulses by changing the polarization of the laser pulse from circular to almost linear, while maintaining a constant intensity of the laser field [17].

In closing this Section, let us note that the generation of attosecond electron wave packets has also been investigated by Varró and Farkas [25]. However, the wave packets analyzed there have a different origin and occur under different circumstances. Namely, their train of pulses result from the interference of electron plane-waves and appear within both plane-wave laser fields and solids. In contrast, our solitary short electron pulses are built from free-electron states. Furthermore, by accounting for the interference effects, the broad structure present in Fig. 5 (upper-right panel) constitute in fact a train of attosecond (or even zeptoseconds) electron pulses observed at the mesoscopic distance $d \approx 53$ μm from the parent ion.

5. Conclusions

In this paper we have analyzed the high-energy photoionization of hydrogen-like ions driven by short and circularly-polarized laser pulses. As it was shown in Refs. [12, 14, 15], the energy spectra of photoelectrons can present interference-dominated or interference-free structures. As we have checked here, those structures present maxima located at photoelectron kinetic energies given by Eq. (10), which is valid provided that the target ions are light enough and the electrons are fast [16, 17]. Furthermore, we have shown in this paper that the central kinetic energies of those structures increase linearly with the intensity of the driving field. Finally, we have demonstrated in Sec. 4 that the interference-dominated patterns lead to the creation of double-hump electron wave packets which appear, however, at very short distances from the parent ion. At mesoscopic distances, on the other hand, pronounced interferences are established. In contrast, the supercontinuum leads to the creation of solitary coherent and short electron pulses, the duration of which can be controlled by modifying the intensity of the laser field.

Acknowledgments

This work is supported by the National Science Centre (Poland) under Grant No. 2014/15/B/ST2/02203. We would like to thank Professor S. Varró for the stimulating discussions at the conference and for drawing our attention to Ref. [25].

References

[1] Zewail A H 2005 *Phil. Trans. R. Soc.* A **363** 315
[2] Lobastov V A, Srinivasan R and Zewail A H 2005 *Proc. Natl. Acad. Sci. U. S. A.* **102** 7069
[3] Baum P and Zewail A H 2009 *Chem. Phys.* **366** 2
[4] Zewail A H 2010 *Phil. Trans. R. Soc.* A **368** 1191
[5] Srinivasan R, Lobastov V A, Ruan C-Y and Zewail A H 2003 *Helv. Chim. Acta* **86** 1761
[6] Zewail A H 2010 *Science* **328** 187
[7] Sciaini G and Miller R J D 2011 *Rep. Prog. Phys.* **74** 096101
[8] Ischenko A A, Aseyev S A, Bagratashvili V N, Panchenko V Ya and Ryabov E A 2014 *Phys.-Usp.* **57** 633
[9] Baum P and Zewail A H 2007 *Proc. Natl. Acad. Sci. U. S. A.* **104** 18409
[10] Aidelsburger M, Kirchner F O, Krausz F and Baum P 2010 *Proc. Natl. Acad. Sci. U. S. A.* **107** 19714
[11] Baum P 2013 *Chem. Phys.* **423** 55
[12] Krajewska K and Kamiński J Z 2016 *Phys. Rev.* A **94** 013402
[13] Cajiao Vélez F, Kamiński J Z and Krajewska K 2017 *J. Phys. B: At. Mol. Opt. Phys.* https://doi.org/10.1088/1361-6455/aaa056
[14] Cajiao Vélez F, Kamiński J Z and Krajewska K 2017 *J. Phys.: Conf. Ser.* **826** 012010
[15] Krajewska K, Cajiao Vélez F and Kamiński J Z 2017 *Proc. of SPIE* **10241** 102411J
[16] Kamiński J Z, Cajiao Vélez F and Krajewska K 2017 *Laser Phys. Lett.* **14** 075301
[17] Krajewska K, Cajiao Vélez F and Kamiński J Z 2017 *EPL* **119** 13001
[18] Reiss H R 1990 *Phys. Rev.* A **42** 1476
[19] Reiss H R 1990 *J. Opt. Soc. Am.* B **7** 574
[20] Krajewska K, Kamiński J Z 2016 *Phys. Lett.* A **380** 1247
[21] Wolkow D M 1935 *Z. Phys.* **94** 250
[22] Ehlotzky F, Krajewska K and Kamiński J Z 2009 *Rep. Prog. Phys.* **72** 046401
[23] Di Piazza A, Müller C, Hatsagortsyan K Z and Keitel C H 2012 *Rev. Mod. Phys.* **84** 1177
[24] Krajewska K and Kamiński J Z 2015 *Phys. Rev.* A **92** 043419
[25] Varró S and Farkas Gy 2008 *Laser Part. Beams* **26** 9

IOP Conf. Series: Journal of Physics: Conf. Series **999** (2018) 012006 doi:10.1088/1742-6596/999/1/012006

Classical plasma dynamics of Mie-oscillations in atomic clusters

H.-J. Kull and A. El-Khawaldeh

Institute for Theory of Statistical Physics, Laser Physics Group, RWTH Aachen University, 52056 Aachen, Germany

E-mail: kull@ilt-extern.fraunhofer.de

Abstract. Mie plasmons are of basic importance for the absorption of laser light by atomic clusters. In this work we first review the classical Rayleigh-theory of a dielectric sphere in an external electric field and Thomson's plum-pudding model applied to atomic clusters. Both approaches allow for elementary discussions of Mie oscillations, however, they also indicate deficiencies in describing the damping mechanisms by electrons crossing the cluster surface. Nonlinear oscillator models have been widely studied to gain an understanding of damping and absorption by outer ionization of the cluster. In the present work, we attempt to address the issue of plasmon relaxation in atomic clusters in more detail based on classical particle simulations. In particular, we wish to study the role of thermal motion on plasmon relaxation, thereby extending nonlinear models of collective single-electron motion. Our simulations are particularly adopted to the regime of classical kinetics in weakly coupled plasmas and to cluster sizes extending the Debye-screening length. It will be illustrated how surface scattering leads to the relaxation of Mie oscillations in the presence of thermal motion and of electron spill-out at the cluster surface. This work is intended to give, from a classical perspective, further insight into recent work on plasmon relaxation in quantum plasmas [1].

1. Introduction

Mie oscillations in atomic clusters are of basic interest in the study of finite-size electronic properties of nanoparticles. The size dependence of the damping constant of Mie oscillations has received much attention and was treated by classical mean free path models [2, 3] and quantum theoretical linear response [4] and random phase approximation calculations [5]. More generally, damping is attributed to collisions, outer ionization [6, 7], surface scattering [16] and radiation emission. In laser excitations damping is often accompanied by linear [8, 9, 10] and nonlinear [11, 12, 13] resonance absorption. In this work we wish to investigate in more detail Mie oscillations and their relaxation by classical particle simulations. For this purpose, the model of a homogeneously charged ion sphere neutralized by N electrons with a soft-core Coulomb interaction is considered. This model has only two dimensionless parameters, the particle number N and the plasma coupling constant Γ. In this work, we consider a weakly coupled plasma with $\Gamma = 0.1$ and clusters in the range of $N = 8000 - 27000$ particles. After a brief introduction to Mie oscillations of clusters and surface scattering within a hollow sphere, the particle simulations are described and results for the size-dependent damping rates are discussed and compared with previous work.

IOP Conf. Series: Journal of Physics: Conf. Series **999** (2018) 012006 doi:10.1088/1742-6596/999/1/012006

2. Mie-oscillations

2.1. Rayleigh theory

We first review some basic properties of Mie oscillations in atomic clusters. Mie oscillations arise from the polarization of the cluster in an external radiation field. The famous Mie theory treats light scattering by a dielectric sphere of an arbitrary size. If the radius of the sphere is much smaller than the wavelength of the radiation, one can restrict attention to the simpler theory of Rayleigh scattering. As originally demonstrated by Rayleigh, a small sphere with dielectric constant ϵ placed in a uniform electric field $\boldsymbol{E}_0$ acquires a homogeneous polarization ([14], p.32, eq.(20))

$$\boldsymbol{P} = \frac{3}{4\pi}\frac{\epsilon-1}{\epsilon+2}\boldsymbol{E}_0. \tag{1}$$

The electric field produced by a homogeneously polarized sphere is given by

$$\boldsymbol{E}_P = -\frac{4\pi}{3}\boldsymbol{P} = -\frac{\epsilon-1}{\epsilon+2}\boldsymbol{E}_0 \tag{2}$$

and the total electric field inside the sphere, consisting of the polarization field and the external field, becomes

$$\boldsymbol{E} = \boldsymbol{E}_0 + \boldsymbol{E}_P = \frac{3}{\epsilon+2}\boldsymbol{E}_0. \tag{3}$$

The electrostatic treatment is still applicable to harmonic fields of frequency ω, if the static dielectric constant is replaced by a frequency-dependent function $\epsilon = \epsilon(\omega)$ and the radius of the sphere is much smaller than the wavelength. The Mie resonance is defined by the condition that the denominator of the electric field (3) becomes zero,

$$\epsilon(\omega) + 2 = 0. \tag{4}$$

At this stage, the Rayleigh theory is purely macroscopic and the Mie-resonance frequencies are just zeros of $\epsilon(\omega) + 2$ for any dielectric function under consideration. As a simple microscopic model, one often chooses a free electron gas with the dielectric function

$$\epsilon(\omega) = 1 - \frac{\omega_p^2}{\omega^2}, \qquad \omega_p^2 = \frac{4\pi e^2 n_0}{m}, \tag{5}$$

where ω_p denotes the plasma frequency, n_0 a homogeneous electron density, e the elementary charge in Gaussian cgs units and m the electron mass. Then the resonance occurs at the well-known Mie frequency

$$\omega_M = \frac{1}{\sqrt{3}}\omega_p. \tag{6}$$

2.2. Thomson model

We now discuss a microscopic model of Mie oscillations in clusters, that is based on the motion of N electrons inside a sphere of radius R with a homogeneous positive charge density en_0. This model is reminiscent of the famous plum-pudding model of Thomson [15], viewing atoms as an extended homogeneous positively charged nucleus and the electrons as point charges immersed into this background. While Thomson's model failed to explain the structure of atoms, it is commonly used as a model of an extended plasma with a neutralizing homogeneous background. The electric field of the homogeneously charged ion sphere with charge density en_0 at a position $\boldsymbol{r}$ inside the sphere is given by

$$\boldsymbol{E}_I(\boldsymbol{r}) = \frac{4\pi}{3}en_0\boldsymbol{r} = \frac{m}{e}\omega_M^2\boldsymbol{r}. \tag{7}$$

IOP Conf. Series: Journal of Physics: Conf. Series **999** (2018) 012006 doi:10.1088/1742-6596/999/1/012006

Inside the sphere, the motion of electron i is described by

$$m\ddot{\boldsymbol{r}}_i = -e\boldsymbol{E}_I(\boldsymbol{r}_i) + \sum_{j\neq i} \frac{e^2}{|\boldsymbol{r}_i - \boldsymbol{r}_j|^3}(\boldsymbol{r}_i - \boldsymbol{r}_j) - e\boldsymbol{E}_0(t), \qquad i,j = 1,\cdots,N \tag{8}$$

where the first term describes the electron-ion, the second one the electron-electron interaction and the last one an external time-dependent field. Here and in the following parts of this section it is assumed, that electrons move inside the sphere without reaching or crossing its boundary. The equation of motion becomes particularly simple, if one considers the center of mass (c.m.)

$$\boldsymbol{R} = \frac{1}{N}\sum_i \boldsymbol{r}_i \tag{9}$$

of all electrons. In this case, (8) has to be averaged over all electrons. For the average of the force one can draw three important conclusions:

(i) Due to the action-reaction law, the electron-electron interactions cancel. The electrons therefore exert no force on the c.m.

(ii) For an isotropic electron distribution, the electron-ion interaction will also vanish for any central field $\boldsymbol{E}_I = -\nabla\phi(r)$ that can be derived from an arbitrary spherically symmetric potential $\phi(r)$. Accordingly, in equilibrium, the c.m. will be assumed force-free and coincident with the center of the ion sphere.

(iii) The average force becomes particularly simple for spatially constant and linear forces. Assuming a force $\boldsymbol{f}(\boldsymbol{r}) = -k\boldsymbol{r} + \boldsymbol{f}_0$ with constants k and $\boldsymbol{f}_0$ acting on the particles at positions $\boldsymbol{r}_i$, the ensemble average will be given by

$$\boldsymbol{F} = \frac{1}{N}\sum_i \boldsymbol{f}(\boldsymbol{r}_i) = -k\boldsymbol{R} + \boldsymbol{f}_0 = \boldsymbol{f}(\boldsymbol{R}).$$

As in the quantum-mechanical Ehrenfest-theorem, the average can be moved in this special case from the force to the positions.

Using these properties for the average of (8), the c.m. obeys the equation of a harmonic oscillator,

$$\ddot{\boldsymbol{R}} + \omega_M^2\boldsymbol{R} = \frac{q}{m}\boldsymbol{E}_0(t). \tag{10}$$

Looking at the driven solution in an external field of frequency ω, one immediately arrives at the induced dipole moment $\boldsymbol{d} = -Ne\boldsymbol{R} = \alpha(\omega)\boldsymbol{E}_0$ and the corresponding polarizability

$$\alpha(\omega) = \frac{Ne^2/m}{\omega_M^2 - \omega^2} \tag{11}$$

of the cluster. As in the Rayleigh theory, the Mie resonance occurs at the Mie frequency ω_M.

2.3. *Vlasov and fluid models*

Another closely related approach treats the electrons of a cluster not as point-particles but in the framework of the kinetic theory by a single-electron distribution function $f(\boldsymbol{r},\boldsymbol{v},t)$ in phase space. The basic kinetic equation for collisionless plasmas is the Vlasov equation

$$\partial_t f(\boldsymbol{r},\boldsymbol{v},t) + \boldsymbol{v}\cdot\partial_{\boldsymbol{r}} f(\boldsymbol{r},\boldsymbol{v},t) - \frac{e}{m}\boldsymbol{E}(\boldsymbol{r},t)\cdot\partial_{\boldsymbol{v}} f(\boldsymbol{r},\boldsymbol{v},t) = 0, \tag{12}$$

where the electric field $\boldsymbol{E}(\boldsymbol{r},t)$ is the sum of the fields produced by the electrons, ions and by the external laser field,

$$\boldsymbol{E} = \boldsymbol{E}_I + \boldsymbol{E}_E + \boldsymbol{E}_0. \tag{13}$$

The first two velocity moments of the distribution function define the particle density n and the mean velocity $\boldsymbol{u}$ according to

$$n = \int d^3v\ f(\boldsymbol{r},\boldsymbol{v},t), \qquad \boldsymbol{u} = \frac{1}{n}\int d^3v\ \boldsymbol{v}f(\boldsymbol{r},\boldsymbol{v},t). \tag{14}$$

Taking the corresponding moments of the Vlasov equation (12) leads in a standard manner to the macroscopic fluid equations

$$\begin{aligned} \partial_t n &+ \nabla\cdot(n\boldsymbol{u}) = 0, & (15a)\\ \partial_t(mn\boldsymbol{u}) &+ \nabla\cdot(mn\boldsymbol{u}\boldsymbol{u}) = -en\boldsymbol{E} - \nabla\cdot\boldsymbol{P} & (15b)\end{aligned}$$

with the pressure tensor

$$\boldsymbol{P} = m\int d^3v'\ \boldsymbol{v}'\boldsymbol{v}'f(\boldsymbol{r},\boldsymbol{v},t), \qquad \boldsymbol{v}' = \boldsymbol{v} - \boldsymbol{u}.$$

The equation of motion of the c.m. can be obtained by integrating the momentum equation (15b) over the volume of a concentric sphere surrounding the ion sphere and enclosing all electrons. The particle number and the c.m. are defined by

$$N = \int d^3r\ n, \qquad \boldsymbol{R}(t) = \frac{1}{N}\int d^3r\ n(\boldsymbol{r},t)\boldsymbol{r}. \tag{16}$$

The c.m. velocity can be expressed with the continuity equation (15a) by the fluid velocity

$$\dot{\boldsymbol{R}} = \frac{1}{N}\int d^3r\ \partial_t n\ \boldsymbol{r} = -\frac{1}{N}\int d^3r\ \nabla\cdot(n\boldsymbol{u})\ \boldsymbol{r} = \frac{1}{N}\int d^3r\ n\boldsymbol{u}. \tag{17}$$

The c.m. acceleration therefore is given by

$$\ddot{\boldsymbol{R}} = \frac{1}{N}\int d^3r\ \partial_t(n\boldsymbol{u}) = -\frac{1}{N}\int d^3r\ \frac{e}{m}n\boldsymbol{E} - \nabla\cdot(n\boldsymbol{u}\boldsymbol{u} + \frac{1}{m}\boldsymbol{P}). \tag{18}$$

The divergence term can be transformed by Gauß's law into a surface integral, which vanishes since the fluid variables vanish by assumption on the surface of the integration volume. Using again the actio-reactio-law and assuming further that all electrons stay inside the ion sphere, the average acceleration is given by

$$\ddot{\boldsymbol{R}} = -\frac{e}{Nm}\int d^3r\ n(\boldsymbol{E}_I(\boldsymbol{r}) + \boldsymbol{E}_0(t)) = -\omega_M^2\boldsymbol{R} - \frac{e}{m}\boldsymbol{E}_0(t). \tag{19}$$

There follows again the harmonic oscillator equation (10). From this derivation it follows that the presence of the kinematic and static pressure in the fluid equations does not affect the c.m. motion. Neither do electron-electron collisions in the point-particle model contribute to the c.m. motion. These conclusions are of course dependent on the neglect of surface effects. The Rayleigh and Thomson models are quite general, however, they severely neglect the electron-ion interaction at and beyond the cluster surface. In the following section, we present a simple model of surface scattering that can demonstrate the origin of an additional friction force due to reflections from the surface of the sphere.

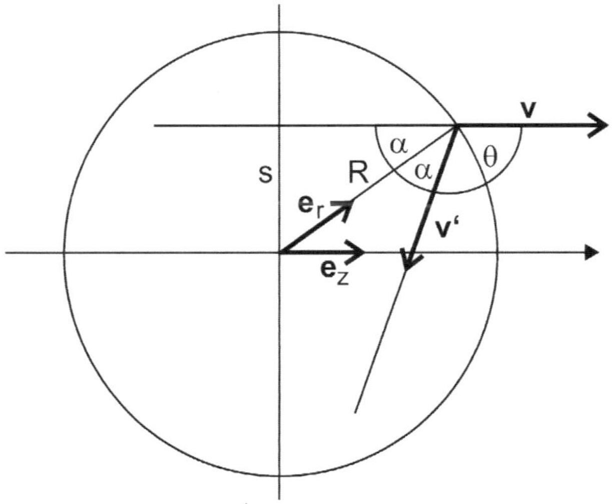

Figure 1. Scattering of a particle at a perfectly reflecting hollow sphere of radius R. The particle is incident with velocity $\boldsymbol{v} = v\boldsymbol{e}_z$ along the z-direction with impact parameter s along a perpendicular direction. The angles of incidence and of reflection are both α. The reflected particle has the velocity $\boldsymbol{v}'$ with $v' = v$, being deflected by an angle θ.

2.4. Hollow-sphere model of surface scattering

The relaxation of Mie oscillations can often be attributed to electron-ion collisions or to collisionless Landau-damping. For larger clusters radiation damping becomes also important. Another cluster-specific relaxation mechanism is due to surface scattering. In this section, we introduce surface scattering by a billiard model. If all electrons remain bounded to the volume of the cluster, one can assume for simplicity an infinitely high potential barrier at the surface of the sphere. We therefore consider a hollow sphere with a perfectly reflecting boundary and derive the friction force due to collisions of free electrons with the cluster surface.

We first assume a beam of particles with a homogeneous density n_0 and a velocity $\boldsymbol{v} = v\boldsymbol{e}_z$ taken along the z-direction. This beam is incident on the inner surface of a hollow sphere as shown in Fig.1. Looking at the reflection of a particle with angle of incidence α, one obtains geometrically the relationship between the impact parameter s and the angle of deflection θ,

$$\theta = \pi - 2\alpha, \qquad s = R\sin\alpha = R\cos(\theta/2). \tag{20}$$

It then follows immediately that the differential cross-section is given by

$$\frac{d\sigma}{d\Omega} = \frac{s}{\sin\theta}\left|\frac{ds}{d\theta}\right| = \frac{1}{4}R^2. \tag{21}$$

It is equal to the more familiar differential cross section for scattering at the outside surface of a hard sphere. The total cross-section becomes $\sigma_T = \pi R^2$ and corresponds to the maximum circular cross-section of the sphere.

The rate of momentum change due to surface collisions can be calculated in terms of the momentum transfer $\Delta\boldsymbol{p}$ for a single collision and the scattering rate $jd\sigma$ as

$$\frac{\Delta\boldsymbol{p}}{\Delta t} = \int \Delta\boldsymbol{p}(s,\phi) j d\sigma = \int d\Omega\, \Delta\boldsymbol{p}(\theta,\phi)\, j\frac{d\sigma}{d\Omega}, \qquad j = n_0 v. \tag{22}$$

During the reflection at the surface, the normal component of the momentum becomes reversed. Writing the momenta $\boldsymbol{p}$ of the incident and $\boldsymbol{p}'$ of the reflected particles as a sum of outward directed normal (r) and tangential (t) components, one obtains

$$\boldsymbol{p} = \boldsymbol{p}_t + \boldsymbol{p}_r, \qquad \boldsymbol{p}' = \boldsymbol{p}_t - \boldsymbol{p}_r, \qquad \Delta\boldsymbol{p} = -2\boldsymbol{p}_r = -2\boldsymbol{p}\cdot\boldsymbol{e}_r\boldsymbol{e}_r. \tag{23}$$

In spherical coordinates the angle dependence of the momentum change can be expressed as

$$\Delta\boldsymbol{p} = -2mv\cos\alpha\begin{pmatrix}\sin\alpha\cos\phi\\ \sin\alpha\sin\phi\\ \cos\alpha\end{pmatrix} = -mv\begin{pmatrix}\sin\theta\cos\phi\\ \sin\theta\sin\phi\\ 1-\cos\theta\end{pmatrix}. \tag{24}$$

Integrating now (24) over the solid angle, all angle-dependent parts average to zero. The only nonvanishing contribution comes from the constant term in the z-component. It yields

$$\frac{\Delta\boldsymbol{p}}{\Delta t} = j\frac{d\sigma}{d\Omega}\int d\Omega\,\Delta\boldsymbol{p} = -n_0 v\sigma_T\, m\boldsymbol{v}. \tag{25}$$

The average rate of momentum change per particle is a velocity dependent friction force

$$\boldsymbol{F}(\boldsymbol{v}) = \frac{1}{N}\frac{\Delta\boldsymbol{p}}{\Delta t} = -\nu(v)m\boldsymbol{v}, \qquad \nu(v) = \frac{3}{4}\frac{v}{R}. \tag{26}$$

Note that the friction coefficient $\nu(v)$ is velocity dependent and that it scales with $1/R$ with the cluster radius. This scaling arises basically from the surface to volume ratio of $\sigma_T/N \propto 1/R$.

Next, we consider an equilibrium with a spatially homogeneous electron density and an arbitrary isotropic velocity distribution $f(\boldsymbol{v}) = n_0 g(v)$ inside the hollow sphere. The distribution function $f(\boldsymbol{v})$ is normalized to the number of particles N within the sphere and therefore

$$\int d^3v\; g(v) = 1. \tag{27}$$

For example, a Fermi distribution has equal weights of all occupied velocity points and is given by

$$g(v) = \frac{3}{4\pi v_F^3}\theta(v_F - v), \tag{28}$$

where the maximum velocity v_F denotes the Fermi-velocity.

The total friction force due to surface reflections then is given by an integral over velocity space,

$$\langle\boldsymbol{F}(\boldsymbol{v})\rangle \equiv \int d^3v\; g(v)\boldsymbol{F}(\boldsymbol{v}). \tag{29}$$

Since $g(v)\boldsymbol{F}(\boldsymbol{v})$ is a central field, the velocity integral vanishes and the friction force is zero. Exciting now a Mie oscillation with velocity $\dot{\boldsymbol{R}}(t)$ inside the cluster, all electrons will be shifted to the new velocity $\boldsymbol{v}' = \boldsymbol{v} + \dot{\boldsymbol{R}}$ and they experience the friction force $\boldsymbol{F}(\boldsymbol{v}')$. Integrating now over all particles yields a nonvanishing average

$$\boldsymbol{F}_M = \langle\boldsymbol{F}(\boldsymbol{v}')\rangle = -\frac{3m}{4R}\langle|\boldsymbol{v}+\dot{\boldsymbol{R}}|(\boldsymbol{v}+\dot{\boldsymbol{R}})\rangle. \tag{30}$$

Since the Mie oscillation is considered a small perturbation, the average is only calculated to first order in $\dot{\boldsymbol{R}}$. Setting $\boldsymbol{v} = v\boldsymbol{n}$ and $\dot{\boldsymbol{R}} = \dot{R}\boldsymbol{e}_z$ one has to first order $v' = v + \boldsymbol{n}\cdot\boldsymbol{e}_z\dot{R}$ and

$$v'\boldsymbol{v}' = v\boldsymbol{v} + v\boldsymbol{n}\cdot\dot{\boldsymbol{R}} + v\dot{\boldsymbol{R}} = v^2\boldsymbol{n} + \boldsymbol{n}\boldsymbol{n}\cdot\boldsymbol{e}_z v\dot{R} + v\dot{\boldsymbol{R}}\,. \tag{31}$$

IOP Conf. Series: Journal of Physics: Conf. Series **999** (2018) 012006 doi:10.1088/1742-6596/999/1/012006

Integrating over the solid angle in velocity space yields

$$\int d\Omega \; \boldsymbol{n} = 0, \qquad \int d\Omega \; \boldsymbol{n}\boldsymbol{n} = \frac{4\pi}{3}\boldsymbol{I}, \tag{32}$$

where $\boldsymbol{I}$ denotes the unit tensor. Using these properties, the remaining integral over the radial coordinate v becomes

$$\langle v'\boldsymbol{v}'\rangle = \frac{16\pi}{3}\dot{\boldsymbol{R}} \int_0^\infty dv \; v^3 g(v). \tag{33}$$

Finally, choosing the Fermi distribution (28), the velocity average becomes $\langle v'\boldsymbol{v}'\rangle = v_F \dot{\boldsymbol{R}}$ and the friction force (30) is given by

$$\boldsymbol{F}_M = -\nu m \dot{\boldsymbol{R}}, \qquad \nu = \frac{3}{4}\frac{v_F}{R}. \tag{34}$$

Comparing with the single-beam result (26), one can see that the force is now linear in the velocity perturbation $\dot{\boldsymbol{R}}$ and that the friction coefficient is evaluated at the Fermi-velocity. Adding this surface-friction force to the center of mass motion yields the damped harmonic oscillator equation

$$\ddot{\boldsymbol{R}} + \nu\dot{\boldsymbol{R}} + \omega_M^2 \boldsymbol{R} = -\frac{e}{m}\boldsymbol{E}_0(t). \tag{35}$$

The decay of an initial perturbation can be defined by the solution to the homogeneous equation ($\boldsymbol{E}_0 = 0$). Assuming an exponential time-dependence $\propto e^{-i\omega t}$ yields complex eigenfrequencies

$$\omega_{1,2} = \pm\sqrt{\omega_M^2 - \left(\frac{\nu}{2}\right)^2} - i\frac{\nu}{2}. \tag{36}$$

and for $\omega_M > \nu/2$ a decay rate of

$$\gamma = \frac{\nu}{2} = A\,\frac{v_F}{2R}, \qquad A = \frac{3}{4}. \tag{37}$$

The exponential decay rate corresponds to a Lorentzian line shape with a full width at half maximum (FWHM) of $\Delta\omega = 2\gamma$.

The surface plasmon decay rate was considered by a number of authors under various assumptions and with slightly different results. A classical approach has been based on the estimate of the mean free path L within finite-size particles. For a sphere, the authors obtain $L = R$ for isotropic [2] and $L = \frac{4}{3}R$ for diffuse scattering [17, 3]. The effective collision frequency was then estimated as

$$\nu = v_F/L = A\frac{v_F}{R} \tag{38}$$

with $A = 1$ or $A = 3/4$, respectively. The present result (37) shows the same scaling and even agrees with the $L = \frac{4}{3}R$ model. However, one should notice that the average mean free path is determined geometrically, while the present scattering model is a kinetic approach based on particle dynamics, specular reflection and the velocity distribution function. Plasmon damping in metal particles was also considered quantum theoretically [4, 18, 19, 5]. In a thorough review of the early work, it was concluded that [5]

$$A = p\,\sqrt{1 + \frac{\omega}{E_F}}, \tag{39}$$

where $p = 3/4$ in quasi-classical and $p = 3/\pi^2$ in random-phase approximation (RPA). The dependence on the excitation frequency ω derives from resonances with quantum-mechanical

IOP Conf. Series: Journal of Physics: Conf. Series **999** (2018) 012006 doi:10.1088/1742-6596/999/1/012006

excitation energies of various dipole transitions. Because of this quantum correction, the quasi-classical result is somewhat larger than the classical estimate (37). The more complete RPA treatment leads to a smaller coefficient. For silver particles ($E_F \approx 5.5$ eV, $\omega \approx 3$ eV) the frequency-dependent factor is about 1.25. In this case the FWHM line widths become $\Delta\omega = 0.94\ v_F/R$ for the semi-classical and $\Delta\omega = 0.38\ v_F/R$ for the RPA case. Both approximations proved to be well consistent with different experimental results [17, 20, 5]. In later experiments, higher absorption values have been attributed to embedding or substrate materials while a smaller value of $A = 0.25$ was measured for free isolated Ag [16] and Au [21, 22] clusters. The classical approach (37) yields a somewhat larger value $A = 0.75$, which may be viewed as a reasonable order of magnitude estimate. A discussion of surface damping under various conditions can be found in [16, 23, 24].

3. Particle simulations

3.1. Computational model

To study the surface damping effects in more detail, we have performed particle simulations of N electrons interacting with a fixed homogeneous ion sphere. The ion sphere has radius R and charge Ne, where e denotes the positive elementary charge. The average electron density is $n_0 = 3N/(4\pi R^3)$ and the mean particle distance r_s can be defined by $4\pi n_0 r_s^3/3 = 1$. It is convenient to introduce dimensionless positions $\boldsymbol{r}' = \boldsymbol{r}/r_s$, times $t' = \omega_M t$ and electric fields $\boldsymbol{E}' = e\boldsymbol{E}/(m\omega_M^2 r_s)$, where ω_M is the Mie frequency (6). Omitting the prime for simplicity of notation, the dimensionless equations of motion for an electron with charge $q = -e$ becomes,

$$\ddot{\boldsymbol{r}}_i = -\boldsymbol{E}_0(t) - \boldsymbol{E}_I(\boldsymbol{r}_i) - \sum_{j\neq i} \boldsymbol{E}_E(\boldsymbol{r}_i - \boldsymbol{r}_j), \qquad i = 1, \cdots, N. \tag{40}$$

Here $\boldsymbol{E}_0(t)$ denotes an external laser field and $\boldsymbol{E}_I$ the electric field of the ion sphere that is given by

$$\boldsymbol{E}_I(\boldsymbol{r}) = \boldsymbol{r}\left(\theta(R-r) + \theta(r-R)\frac{R^3}{r^3}\right). \tag{41}$$

The electron-electron interaction $\boldsymbol{E}_E$ will be modelled by a softcore Coulomb potential and a corresponding electric field,

$$\phi(\boldsymbol{r}) = -\frac{1}{\sqrt{r^2+\epsilon^2}}, \qquad \boldsymbol{E}_E(\boldsymbol{r}) = -\nabla\phi(\boldsymbol{r}) = -\frac{\boldsymbol{r}}{(r^2+\epsilon^2)^{3/2}}. \tag{42}$$

A constant $\epsilon = 2$ is used in the simulations to eliminate hard collisions with impact parameters near and below the interparticle distance r_s. Apart from this constant, the equations of motion (40) contain only a single parameter, which is the dimensionless radius $R = N^{1/3}$ of the sphere.

The equations of motion have to be supplemented by initial conditions. Initially, the electrons are placed at random positions inside the sphere. The velocities are chosen with random directions and the kinetic energies are distributed according to the Fermi distribution (28). The Fermi energy E_F defines another dimesionless parameter of the model, which is the coupling parameter $\Gamma = e^2/(r_s E_F))$ of the plasma. In the present calculations, we have always chosen a weakly coupled plasma with $\Gamma = 0.1$. The radius R is varied in the range $20 - 30$, corresponding to particle numbers between 8000 and 27000.

3.2. Results

We now discuss some of the results of the present particle simulations for a cluster with dimensionless radius $R = 28$ and coupling parameter $\Gamma = 0.1$. To study the evolution of the Mie plasmon and its damping, the electron ensemble is initially rigidly shifted by a displacement

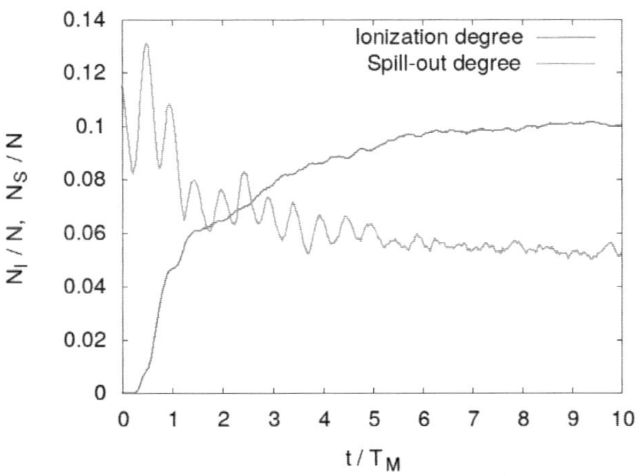

Figure 2. Evolution of the spill-out degree and the outer-ionization degree over 10 Mie periods. In an initial phase of approximately $4-5$ Mie periods, the outer-ionization degree rises and subsequently reaches a quasi-stationary regime. The initial spill-out degree corresponds to the initial shift of the particles with respect to the ion sphere. The spill-out degree initially oscillates with twice the Mie frequency, since electrons are ejected in both maxima of one Mie oscillation. Finally, it also approaches some quasi-stationary regime. ($R = 28$, $\Gamma = 0.1$, $\Delta x = 0.15R$)

$\Delta x = 0.15R$ from the equilibrium configuration. The evolution of the mean velocity of the electrons then indicates exponentially damped Mie oscillations. In contrast to the simplified models of Sec. 2, the electrons can reach and cross the cluster surface. To demonstrate electron spill out and outer ionization of the cluster, we have defined a spill-out degree as the fraction of particles in a shell $R < r < 1.5R$ and an outer-ionization degree as the fraction of particles in the exterior region $r > 1.5R$. Fig. 2 shows the evolution of both parameters over the simulation time of 10 Mie periods $T_M = 2\pi/\omega_M$. It can be seen that electrons are ejected in each half wave of the Mie oscillation, leading finally to an outer-ionization degree of about 10 %. The present simulation model does not allow to lower the ionization degree well below the 1 % level. The present initial ensemble is no perfect equilibrium and one therefore has to accept some outer ionization even without perturbation.

The approach to a quasi-stationary regime can also be recognized in the evolution of the particle energies. In Fig. 3, one can recognize the mean kinetic energy per particle $W = \frac{1}{2}\overline{v^2}$ and the temperature T defined by $\frac{3}{2}T = \frac{1}{2}\overline{(v-\overline{v})^2}$. The initial Fermi energy $E_F = 1/\Gamma = 10$ corresponds to the mean energy $W = (3/5)E_F = 6$ and the temperature $T = \frac{2}{3}W = 4$. While the kinetic energy of the mean velocity decreases, the temperature increases and becomes nearly constant in the quasi-stationary regime.

The evolution of the potential energies of the electrons can be seen in Fig. 4. The electron-ion potential is the energy of all electrons in the potential of the ion sphere per particle, the electron-electron potential is the interaction energy of all electron pairs per particle. We have also convinced that the total energy consisting of the kinetic energy and the potential energies is very well conserved in the calculations up to the accuracy of the finite-difference scheme. The potential energies decrease in magnitude due to the expansion of the electron cloud. Noting the different scales of the axes, this decrease is naturally much larger in the ionization regime shown in (a),(c) than in the final quasi-stationary regime (b),(d). Looking at the oscillation frequencies of the potential energies, one can observe a decrease from $2\omega_M$ to frequencies in the

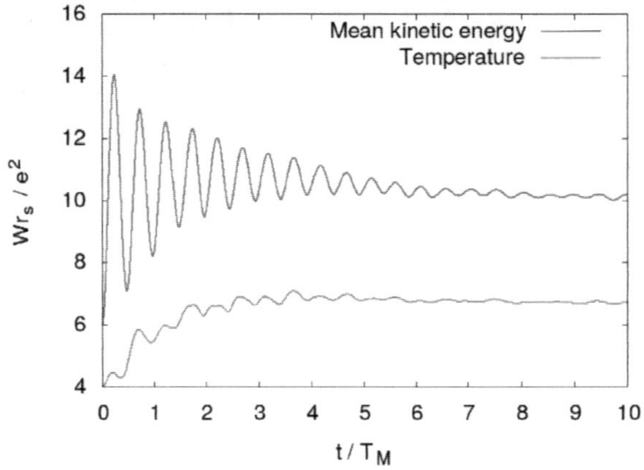

Figure 3. Evolution of the mean kinetic energy and the temperature. In the initial phase, the kinetic energy of the Mie oscillation is gradually converted into temperature. In the final stage of the evolution, no further heating is observed and the temperature remains nearly constant. ($R = 28$, $\Gamma = 0.1$, $\Delta x = 0.15R$)

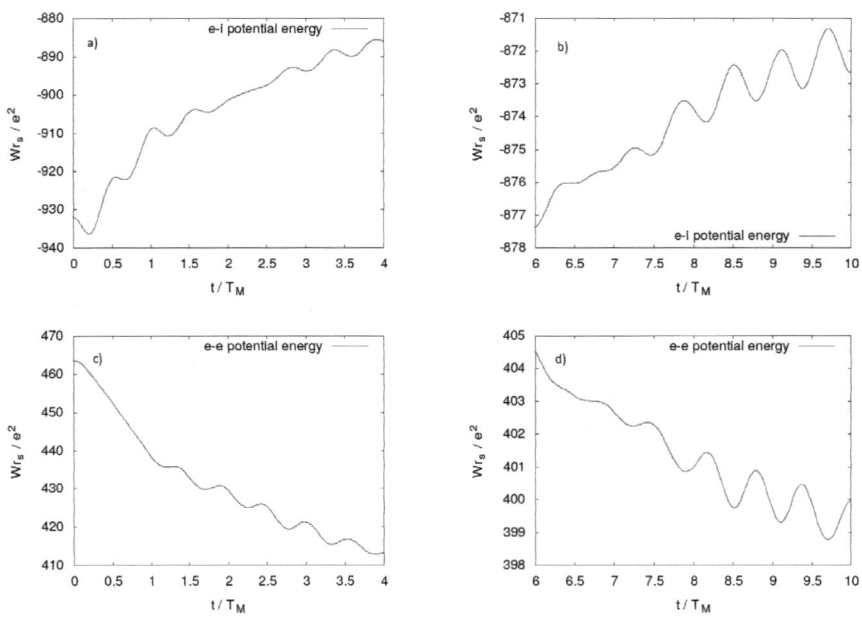

Figure 4. Evolution of potential energies. The magnitude of the potentials decreases due to the increase of the extension of the electron cloud. Furthermore, a decrease of the oscillation frequency indicates a conversion from surface modes with frequency $2\omega_M$ to volume modes with frequencies in the range of ω_p. a) Initial and b) final stage of the electron-ion potential energy. c) Initial and d) final stage of the elelctron-electron potential energy. ($R = 28$, $\Gamma = 0.1$, $\Delta x = 0.15R$)

range of ω_p and opposite phases in the electron-electron and electron-ion potentials. This shift indicates a conversion from surface to volume modes that has been observed in recent quantum calculations of plasmon decay [1]. The phase difference also indicates radial oscillations of the electron density inside the ion sphere where a compression increases the electron repulsion but decreases the ion attraction. Although the mode conversion process has been demonstrated much more convincingly in the quantum regime, the present classical particle simulations can confirm a similar behaviour in the quasi-stationary regime following the ionization stage.

For sufficiently large clusters, we have observed an exponential decay of the Mie oscillations. Using again the preceeding parameters, the oscillation and the decay of the mean velocity V is shown in Fig. 5. The damped oscillation can be very well fitted to the function

$$V(x) = a\sin(\omega t + p)e^{-\gamma t} + c \tag{43}$$

with fit parameters a, ω, p, γ and c. The constant c accounts for a small drift velocity. The fit is applied both to the initial and final phases as indicated in Fig. 5. In this manner, one can define a maximum and a minimum decay rate γ corresponding to the initial ionization phase and the final quasi-stationary phase, respectively. These decay rates have been evaluated as a function of the cluster radius in the range $R = 20$ up to $R = 30$ as shown in Fig. 6. The decay rates follow roughly the expected $1/R$ scaling of (37), however, the deviations from this simple law are appreciable. We attribute the scattering of the data to a mixture of various decay mechanisms including collisional damping, ionization damping, surface damping and damping by mode conversion. The calculated decay rates are well consistent with previous models. Evaluating for instance the damping constant A in (37) for $\tilde{\gamma} = 2\pi\gamma \approx 0.2 - 0.4$, $R = 25$ and with an approximate Fermi energy of $E_F = \frac{5}{3}W \approx 15$ in the final stationary phase, one obtains

$$A = \frac{2R}{v_F}\gamma = \frac{2R}{\sqrt{2E_F}}\frac{\tilde{\gamma}}{2\pi} \approx 0.3 - 0.6. \tag{44}$$

This range of values is close to commonly accepted decay constants as discussed below (39).

4. Conclusions

We have investigated the decay of Mie oscillations in spherical clusters by classical particle simulations. In appropriate units, the physical model contains only two physical parameters namely the cluster radius $R = N^{1/3}$ and the plasma coupling parameter $\Gamma = e^2/(r_s E_F)$. We have restricted attention to the weakly coupled regime, setting $\Gamma = 0.1$. For sufficiently large clusters with radii larger than 20, we observed an exponential decay of the Mie oscillations. The decay rate has been calculated in the range $R = 20$ up to $R = 30$, where a calculation of the electron-electron interaction was computationally feasible. For larger clusters, the computational effort of calculating pairwise particle interactions increases strongly and one has to resort to other methods. The decay constants proved in the expected range of values, although considerable deviations from the simple scaling law (39) have been observed. We conclude that a mixture of different damping processes including outer ionization, collisional damping, surface damping and mode conversion may contribute. Outer ionization could not be completely suppressed most likely due to an imperfect equilibrium state. Collisional damping depends on the parameter ϵ of the soft-core potential. Decreasing its value the damping constant has increased. This shows that one is not in a perfect collisionless regime. Surface damping is likely to play a dominant role since we obtained close quantitative agreement with the corresponding scaling law and have observed the related mode conversion process described in [1]. Plasma-wave dynamics in the interior of the cluster has also been observed in previous particle simulations of laser-excited clusters [25, 26].

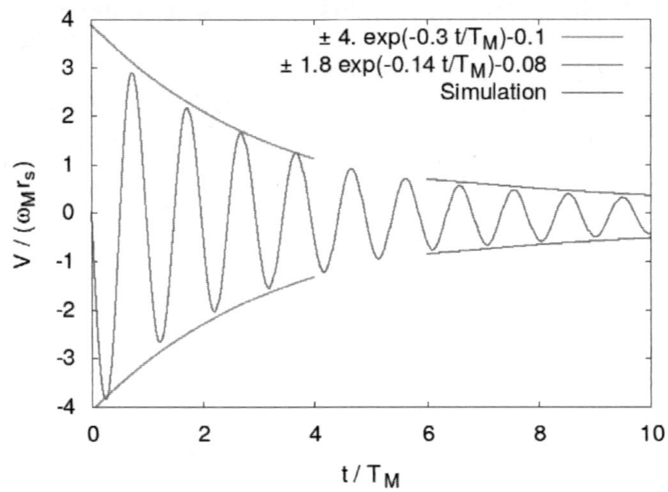

Figure 5. Mie oscillation of the mean velocity V and exponential decay laws in the initial and final stages of the calculation. ($R = 28$, $\Gamma = 0.1$, $\Delta x = 0.15R$)

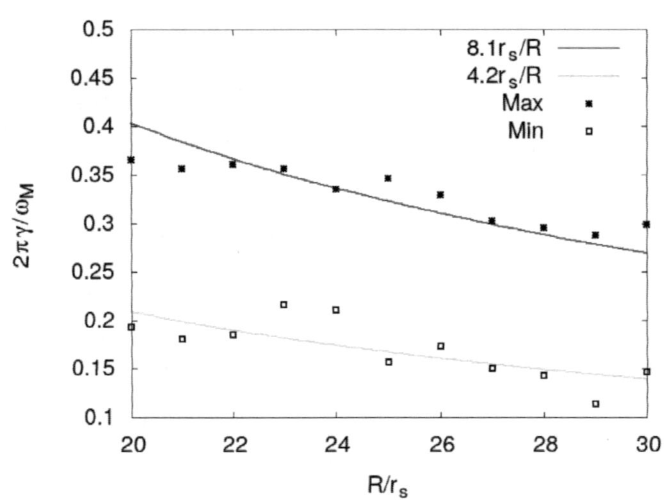

Figure 6. Decay rate γ of Mie oscillations as a function of the cluster radius R. The maximum rate corresponds to the first four periods, the minimum rate to the last four periods considered. Both rates are compared to $1/R$ fitting functions. ($R = 20 - 30$, $\Gamma = 0.1$, $\Delta x = 0.15R$)

IOP Conf. Series: Journal of Physics: Conf. Series **999** (2018) 012006 doi:10.1088/1742-6596/999/1/012006

[1] El-Khawaldeh A and Kull H J 2017 *Phys. Rev. A* **95**(4) 043401
[2] Euler J 1954 *Z. Phys.* **137** 318
[3] Coronado E A and Schatz G C 2003 *J. Chem. Phys.* **119** 3926–3934
[4] Kawabata A and Kubo R 1966 *J. Phys. Soc. Jpn.* **21** 1765
[5] Thoai D T and Ekardt W 1982 *Solid State Commun.* **41** 687 – 690
[6] Greschik F, Arndt L and Kull H J 2005 *EPL (Europhys. Lett.)* **72** 376–382
[7] Petrov G M and Davis J 2006 *Phys. Plasmas* **13** 033106
[8] Ditmire T, Donnelly T, Rubenchik A M, Falcone R W and Perry M D 1996 *Phys. Rev. A* **53** 3379–3402
[9] Döppner T, Fennel T, Diederich T, Tiggesbäumker J and Meiwes-Broer K H 2005 *Phys. Rev. Lett.* **94**(1) 013401
[10] Mikaberidze A, Saalmann U and Rost J M 2008 *Phys. Rev. A* **77**(4) 041201
[11] Fomichev S V, Popruzhenko S V, Zaretsky D F and Becker W 2003 *J. Phys. B* **36** 3817
[12] Kundu M and Bauer D 2006 *Phys. Rev. Lett.* **96**(12) 123401
[13] Mulser P, Bauer D and Ruhl H 2008 *Phys. Rev. Lett.* **101**(22) 225002
[14] Lord Rayleigh FRS 1897 *The London, Edinburgh, and Dublin Philosophical Magazine and Journal of Science* **44:266** 28–52
[15] Thomson J J FRS 1904 *The London, Edinburgh, and Dublin Philosophical Magazine and Journal of Science* **7:39** 237–265
[16] Hövel H, Fritz S, Hilger A, Kreibig U and Vollmer M 1993 *Phys. Rev. B* **48**(24) 18178–18188
[17] Kreibig U 1974 *J. Phys. F: Met. Phys.* **4** 999
[18] Lushnikov A A and Simonov A J 1974 *Z. Phys.* **270** 17–24
[19] Lozovik Y E and Nishanov V N 1978 *Sov. Phys. Solid State* **20(12)** 2111
[20] Abe H, Schulze W and Tesche B 1980 *Chem. Phys.* **47** 95 – 104
[21] Berciaud S, Cognet L, Tamarat P and Lounis B 2005 *Nano Lett.* **5** 515–518
[22] Novo C, aand Jorge Perez-Juste D G, Zhang Z, Petrova H, Reismann M, Mulvaney P and Hartland G V 2006 *Phys. Chem. Chem. Phys.* **8** 3540–3546
[23] Gildenburg V B, Kostin V A and Pavlichenko I A 2011 *Phys. Plasmas* **18** 092101
[24] Juvé V, Cardinal M F, Lombardi A, Crut A, Maioli P, Pérez-Juste J, Liz-Marzán L M, Del Fatti N and Vallée F 2013 *Nano Lett.* **13** 2234–2240
[25] Greschik F and Kull H-J, 2004 *Laser and Particle Beams* **22** 137–145
[26] Varin C, Peltz C, Brabec T and Fennel T. 2012 *Phys. Rev. Lett.* **108** 175007

The Spiral of Life

F Cajiao Vélez, J Z Kamiński and K Krajewska

Institute of Theoretical Physics, Faculty of Physics, University of Warsaw, Pasteura 5, 02-093 Warsaw, Poland

E-mail: `Felipe.Cajiao-Velez@fuw.edu.pl`

Abstract. High-energy photoionization driven by short and circularly-polarized laser pulses is studied in the framework of the relativistic strong-field approximation. The saddle-point analysis of the integrals defining the probability amplitude is used to determine the general properties of the probability distributions. Additionally, an approximate solution to the saddle-point equation is derived. This leads to the concept of the three-dimensional spiral of life in momentum space, around which the ionization probability distribution is maximum. We demonstrate that such spiral is also obtained from a classical treatment.

1. Introduction

The recent development of laser facilities like ELI [1] or XCELS [2], where short laser pulses can reach power of few petawatts, calls for a better understanding of the high-energy photoionization, where relativistic and spin effects may occur. Under such circumstances, the numerical solution of the wave equations becomes computationally demanding due to a very large number of grid points to be taken into account in the temporal and spatial dimensions. For this reason, other approaches are required.

Recently, we have analyzed the high-energy ionization driven by intense, circularly-polarized, few-cycle laser pulses under the framework of the so-called relativistic strong-field approximation (RSFA) [3, 4]. We have demonstrated that, for laser field intensities larger than 10^{17} W/cm^2, the resulting photoelectrons can have kinetic energies ranging from tens of keV up to MeV (or even GeV if the laser light is very powerful) [5, 6, 7, 8]. Also, according to our results, the probability distributions of ionization can present two types of structures: interference-free (the so-called supercontinua) or interference-dominated. With the help of the saddle-point analysis of the probability amplitudes, we have located the regions where such structures appear in the energy spectra of photoelectrons. In Refs. [9, 10] we have proved that it is possible to construct an approximate solution to the saddle-point equation. This solution defines the three-dimensional spiral of life in momentum space and contains most of the information related to the high-energy photoionization process. It is the aim of this paper to further analyze the properties of this spiral from the perspective of the RSFA, and to compare them with the predictions arising from a purely classical treatment based on the solution of the Newton-Lorentz equations of motion.

2. Theory

We consider an initially unperturbed hydrogen-like ion in the presence of a strong and short laser pulse, lasting for time T_p. The relativistic probability amplitude of photoionization, which relates to the transition of an electron from the initial bound state $\Psi_{\rm i}(\boldsymbol{x})$ of energy E_0 to the continuum, is given by

$$\mathcal{A}_{\rm fi} = -{\rm i}\int {\rm d}^4x {\rm e}^{-{\rm i}(E_0/c)x^0}\bar{\Psi}_{\rm f}(x)e\not{A}(x)\Psi_{\rm i}(\boldsymbol{x}). \tag{1}$$

Here, $A(x)$ is the four-vector potential describing the laser pulse and $\Psi_{\rm f}(x)$ is the scattering state of energy $E_{\boldsymbol{p}}$. Even though Eq. (1) is exact, it cannot be used for very intense laser fields. The reason being that, with the current computer capabilities, $\Psi_{\rm f}(x)$ can only be obtained numerically for laser fields of moderate intensities. That is why, in the RSFA [3, 4, 5], the scattering state is expanded with respect to the binding potential in a Born series. If the kinetic energy of the photoelectrons is much larger than the ionization potential of the system, i.e.,

$$\sqrt{(m_{\rm e}c^2)^2 + (c\boldsymbol{p})^2} - m_{\rm e}c^2 \gg m_{\rm e}c^2 - E_0, \tag{2}$$

the relevant term of the above-mentioned expansion is its zeroth-order. To obtain an analytical form of the scattering state in the lowest order Born expansion, the so-called plane-wave-fronted pulse approximation (PWFA) is also required. In this approximation, the laser field propagating with the speed of light c depends on the time and space variables only through a real phase, $\phi = k\cdot x = \omega(t - \boldsymbol{n}\cdot\boldsymbol{x}/c)$. Here, $k = k^0(1,\boldsymbol{n}) \equiv k^0 n$ is the wave four-vector, $\boldsymbol{n}$ is the propagation direction of the laser field, $k^0 = \omega/c$ is the wave number, and $\omega = 2\pi/T_p$ is the fundamental frequency of field oscillations. To be more specific, we will describe the laser field through the following vector potential,

$$A(\phi) = A_0[\varepsilon_1 f_1(\phi) + \varepsilon_2 f_2(\phi)], \tag{3}$$

where $f_i(\phi)$, $i = 1, 2$, are two arbitrary shape functions which vanish for $\phi \leqslant 0$ and $\phi \geqslant 2\pi$. Furthermore, in Eq. (3), $\varepsilon_i = (0, \boldsymbol{\varepsilon}_i)$ are two polarization four vectors such that $\varepsilon_i\cdot\varepsilon_j = -\delta_{ij}$ and $k\cdot\varepsilon_i = 0$.

In the zeroth-order of the Born expansion and using the PWFA, the electron scattering state $\Psi_{\rm f}(x)$ in Eq. (1) is approximated by the so-called Volkov solution [11]; here, denoted as $\psi^{(+)}_{\boldsymbol{p}\lambda}(x)$. It describes the quantum-mechanical behavior of a free electron of asymptotic momentum $\boldsymbol{p}$, the energy $E_{\boldsymbol{p}} = \sqrt{(m_{\rm e}c^2)^2 + (c\boldsymbol{p})^2}$, and the spin degree of freedom $\lambda = \pm$ in the laser pulse (3). Therefore, in the RSFA, the probability amplitude of photoionization can be written as [5, 7, 8]

$$\mathcal{A}_{\lambda_{\rm i}\lambda}(\boldsymbol{p}) = -{\rm i}\int\frac{{\rm d}^3q}{(2\pi)^3}\int {\rm d}^4x\, {\rm e}^{-{\rm i}q\cdot x}\bar{\psi}^{(+)}_{\boldsymbol{p}\lambda}(x)e\not{A}(x)\tilde{\Psi}_{\rm i}(\boldsymbol{q}), \tag{4}$$

where $q = (E_0/c, \boldsymbol{q})$ and $\tilde{\Psi}_{\rm i}(\boldsymbol{q})$ is the Fourier transform of the initial state $\Psi_{\rm i}(\boldsymbol{x})$. Also, $\lambda_{\rm i} = \pm$ is the spin degree of freedom of the initial electron. At this point, we would like to emphasize that the RSFA and, in particular, Eq. (4) are valid provided that the high-energy photoionization is considered, i.e., that the relation (2) holds.

Next, introducing the light cone coordinates (for their definition, see Refs. [12, 13]), most integrals in Eq. (4) can be performed exactly. In doing so, we obtain [10]

$$\mathcal{A}_{\lambda_{\rm i}\lambda}(\boldsymbol{p}) = \mathcal{N}\int_0^{2\pi}{\rm d}\phi\, \exp\left[{\rm i}\int_0^{\phi}{\rm d}u\, G'(u)\right]\bar{B}_{\boldsymbol{p}\lambda}(\phi)\cdot\tilde{\Psi}_{\rm i}(\boldsymbol{Q}), \tag{5}$$

where $\mathcal{N}$ contains all irrelevant factors, $\boldsymbol{Q} = \boldsymbol{p} + (q^0 - p^0)\boldsymbol{n}$, and

$$G'(u) = g_0 + g_1 f_1(u) + g_2 f_2(u) + h\big(f_1^2(u) + f_2^2(u)\big). \tag{6}$$

Here,

$$g_0 = \frac{p^0 - q^0}{k^0},\quad h = \frac{(m_{\rm e}c\mu)^2}{2k\cdot p},\quad g_j = -m_{\rm e}c\mu\frac{\varepsilon_j\cdot p}{k\cdot p}\quad \text{for}\quad j = 1, 2, \tag{7}$$

and $\mu = |eA_0|/(m_e c)$ is the normalized amplitude of the vector potential. Note that, in Eq. (5), the exponential factor oscillates fast as a function of the phase ϕ. This is in contrast to the term $\bar{B}_{\boldsymbol{p}\lambda}(\phi)$, which is slow-oscillating in the interval $0 \leqslant \phi \leqslant 2\pi$. (Even though we do not present here the explicit form of the function $\bar{B}_{\boldsymbol{p}\lambda}(\phi)$, it is given in Eq. (11) of Ref. [10]). This allows us to approximate the integral in (5) using the saddle-point method. In doing so, we obtain that the spin-resolved probability amplitude of ionization, in the RSFA and in the saddle-point approximation, is

$$\mathcal{A}^{\text{saddle}}_{\lambda_i \lambda}(\boldsymbol{p}) = \mathcal{N} \sum_s \mathrm{e}^{\mathrm{i}G(\phi_s)} \sqrt{\frac{2\pi \mathrm{i}}{G''(\phi_s)}} \bar{B}_{\boldsymbol{p}\lambda}(\phi_s) \cdot \tilde{\Psi}_{\mathrm{i}}(\boldsymbol{Q}). \tag{8}$$

The sum in (8) is carried out over all relevant complex saddle points; here, denoted as ϕ_s. The latter are calculated from the equation

$$G'(\phi_s) \equiv \frac{\mathrm{d}G(\phi)}{\mathrm{d}\phi}\bigg|_{\phi_s} = 0, \tag{9}$$

with $\mathrm{Im}\, G(\phi_s) > 0$ or, equivalently, $\mathrm{Im}\, \phi_s > 0$. Also, it is assumed that $\mathrm{Re}\, \phi_s$ determines the time of photoelectron emission, i.e., the moment at which the electron appears in the continuum.

From the analysis of Eq. (8), the following conclusions can be drawn. First, the saddle points contributing the most to the probability amplitude of ionization are those characterized by a positive imaginary part which is as close to zero as possible. Second, if there are two or more of such points in a given range of photoelectron kinetic energies, strong interference effects are expected. Third, if only one saddle point dominates, interference-free structures (supercontinua) should be observed in the energy spectra of photoelectrons. Those properties have been studied in Refs. [5, 7, 8, 9, 10] and are going to be analyzed here in relation to the spiral of life (Sec. 3.1).

Once the probability amplitude of ionization is calculated, the spin-independent probability distribution $\mathcal{P}(\boldsymbol{p})$, in atomic units, can be determined according to (for the details see, Refs [7, 8])

$$\mathcal{P}(\boldsymbol{p}) = \frac{\alpha^2 m_e V E_{\boldsymbol{p}}}{2(2\pi)^3} \sum_{\lambda_i, \lambda = \pm} |\boldsymbol{p}| \cdot |\mathcal{A}_{\lambda_i \lambda}(\boldsymbol{p})|^2. \tag{10}$$

Here, α is the fine structure constant and V is the quantization volume. Note that, as $\mathcal{N}$ in Eq. (5) contains a factor $\sim 1/\sqrt{V}$, $\mathcal{P}(\boldsymbol{p})$ is independent of V. In the numerical analysis presented in this paper, the probability distributions of photoionization will be calculated from Eq. (10) where $\mathcal{A}_{\lambda_i \lambda}(\boldsymbol{p})$ is the exact probability amplitude in the RSFA [Eq. (5)]. The saddle-point analysis and, in particular, Eq. (8) will be only used to interpret those results.

2.1. Spiral of life: Approximate solution to the saddle-point equation

The azimuthal angle φ, i.e., the angle in the plane perpendicular to the laser pulse propagation direction, can be defined in terms of the vector potential $\boldsymbol{A}(\phi)$. Namely,

$$\varphi = \arg\big(e\boldsymbol{A}(\phi) \cdot \boldsymbol{\varepsilon}_1 + \mathrm{i}e\boldsymbol{A}(\phi) \cdot \boldsymbol{\varepsilon}_2\big), \tag{11}$$

where $\arg(z) \in [0, 2\pi[$ is the argument of the complex number z. Conversely, it is possible to define the vector potential as a function of the azimuthal angle, i.e., $\boldsymbol{A}(\varphi) \equiv \boldsymbol{A}(\phi)$. Note, however, that the relation between ϕ and φ is not one-to-one and two or more phases can lead to the same values of φ. This, in turn, means that the curve $\boldsymbol{A}(\varphi)$ can intersect itself at certain azimuthal angles for laser pulses comprising of more than one field oscillation [9, 10]. In order to proceed, we also introduce the time-dependent ponderomotive energy of an electron in the laser field,

$$U_p(\phi) = \frac{e^2 \boldsymbol{A}^2(\phi)}{2m_{\rm e}} = U_p(\varphi). \tag{12}$$

In the same way as the vector potential, the curve $U_p(\varphi)$ can intersect itself.

As it was mentioned before, the phases ϕ_s which possess imaginary parts close to zero are the ones that contribute the most to the sum in (8). Therefore, as a first approximation, we can look for a real phase $\phi_{\boldsymbol{p}}$ which satisfies Eq. (9),

$$G'(\phi_{\boldsymbol{p}}) = \frac{p^0 - q^0}{k^0} - \frac{e\boldsymbol{A}(\phi_{\boldsymbol{p}}) \cdot \boldsymbol{p}}{k \cdot p} + \frac{e^2 \boldsymbol{A}^2(\phi_{\boldsymbol{p}})}{2k \cdot p} = 0. \tag{13}$$

In order to solve this equation, we define the parallel and perpendicular components of the asymptotic momentum $\boldsymbol{p}$ with respect to the propagation direction of the laser field as $p^{\|} = \boldsymbol{p}\cdot\boldsymbol{n}$ and $\boldsymbol{p}^{\perp} = \boldsymbol{p} - p^{\|}\boldsymbol{n}$, respectively. Assuming that [9, 10]

$$\boldsymbol{p}^{\perp} = e\boldsymbol{A}(\phi_{\boldsymbol{p}}), \tag{14}$$

we find out that the saddle-point equation (13) requires

$$p^{\|} = p^0 - q^0 = \frac{e^2 \boldsymbol{A}^2(\phi_{\boldsymbol{p}})}{2q^0}. \tag{15}$$

Hence, we conclude that the photoelectrons with momentum $\boldsymbol{p}(\phi_{\boldsymbol{p}})$ given by

$$\boldsymbol{p}(\phi_{\boldsymbol{p}}) = e\boldsymbol{A}(\phi_{\boldsymbol{p}}) + \frac{e^2 \boldsymbol{A}^2(\phi_{\boldsymbol{p}})}{2q^0}\boldsymbol{n} \tag{16}$$

are detected with maximum probability. This equation defines the *spiral of life in ionization* (also called the momentum spiral). However, it is worth mentioning that the four-momentum associated with $\boldsymbol{p}(\phi_{\boldsymbol{p}})$ is not exactly on the mass shell, as $p\cdot p = E_0^2/c^2$. The ground state energy of a hydrogen-like ion depends on the atomic number Z according to $E_0 = m_{\rm e}c^2\sqrt{1 - Z^2\alpha^2}$ [14]. So, if we have $Z\alpha \ll 1$ (light ions) then $E_0 \approx m_{\rm e}c^2$ and $p\cdot p \approx m_{\rm e}^2c^2$. This reflects the approximate nature of our treatment [9, 10].

Let us now go back to Eq. (14). This equation implies that the azimuthal angle of electron detection, $\varphi_{\boldsymbol{p}}$, is given by

$$\varphi_{\boldsymbol{p}} = \arg\big(e\boldsymbol{A}(\phi_{\boldsymbol{p}}) \cdot \boldsymbol{\varepsilon}_1 + {\rm i}e\boldsymbol{A}(\phi_{\boldsymbol{p}}) \cdot \boldsymbol{\varepsilon}_2\big). \tag{17}$$

This means that, for a given $\varphi_{\boldsymbol{p}}$, one or more phases $\phi_{\boldsymbol{p}}$ are obtained. In this case, Eq. (15) determines the photoelectron energy at which the probability of photoionization is maximum. Namely,

$$E_{\boldsymbol{p}} = E_0 + \frac{c^2 e^2 \boldsymbol{A}^2(\phi_{\boldsymbol{p}})}{2E_0}, \tag{18}$$

which for a light ion takes the form $E_{\boldsymbol{p}} - m_{\rm e}c^2 \approx U_p(\phi_{\boldsymbol{p}})$, where $U_p(\phi)$ is the ponderomotive energy given in (12). Thus, if we interpret $\phi_{\boldsymbol{p}}$ as the phase of the electron birth (real part of the saddle point), we can conclude that the final kinetic energy of the photoelectron is equal to its ponderomotive energy at the moment of appearance in the continuum. This will be analyzed from a classical perspective in Sec. 4.

Finally, from Eq. (15) and for light ions, the polar angle of electron detection, $\theta_{\boldsymbol{p}}$, is given by

$$\cos\theta_{\boldsymbol{p}} = \frac{p^{\|}}{|\boldsymbol{p}|} \approx \sqrt{\frac{E_{\boldsymbol{p}} - m_{\mathrm{e}}c^2}{E_{\boldsymbol{p}} + m_{\mathrm{e}}c^2}}. \tag{19}$$

Note that, for small kinetic energies ($E_{\boldsymbol{p}} \approx m_{\mathrm{e}}c^2$) the polar angle is close to $\pi/2$ and it decreases with increasing $E_{\boldsymbol{p}}$. This is a consequence of the radiation pressure exerted by the laser field on the photoelectrons [5, 15].

In this Section we have shown that the three-dimensional spiral of life (16) determines the maxima of the probability distribution of high-energy photoelectrons. At the given azimuthal angle, the phase $\phi_{\boldsymbol{p}}$ predicts the kinetic energy for which the electrons will be observed with the largest probability. Furthermore, from Eq. (19), the polar angle of detection can be calculated directly. Another interesting feature is that the three-dimensional spiral can also intersect itself. Such self-intersections are characterized by two different phases $\phi_{\boldsymbol{p}}$ (i.e., two contributing saddle-points) leading to the same asymptotic momentum. Therefore, the self-intersections of the curve $\boldsymbol{p}(\phi_{\boldsymbol{p}})$ determine the directions in space and the kinetic energies at which the interference-dominated patterns can be observed. Even though the momentum spiral is a very powerful construction, its predictions are valid provided that the ions are light ($Z\alpha \ll 1$) and the photoelectrons are fast [i.e., if the relation (2) is also fulfilled]. In the next Section, the last statement will be quantified.

3. Numerical calculations

In the following, we consider a hydrogen-like ion with $Z = 2$ (He^+) in the presence of a circularly-polarized, intense, and short laser pulse propagating along the z-direction ($\boldsymbol{n} = \boldsymbol{e}_z$). The pulse comprises of five field oscillations ($N_{\mathrm{osc}} = 5$) within a $\sin^2$ envelope and has a carrier-envelope phase $\chi = \pi$. The explicit form of the corresponding shape functions can be found in Eqs. (40) to (44) of Ref. [5]. For our numerical analysis, we choose the carrier frequency $\omega_{\mathrm{L}} = N_{\mathrm{osc}}\omega$ of a Ti-Sapphire laser field ($\omega_{\mathrm{L}} = 1.5498\,\mathrm{eV}$) and the time-averaged intensity $I = 10^{17}\ \mathrm{W/cm^2}$.

3.1. Probability of ionization along the spiral

As it was mentioned above, the three-dimensional spiral of life determines the momenta for which the probability of ionization is maximal. Therefore, in this Section, the behavior of the saddle points and the probability distributions along the spiral are going to be analyzed.

In the left panel of Fig. 1 we present the three-dimensional spiral $\boldsymbol{p}(\phi_{\boldsymbol{p}})$ [Eq. (16)] for the laser pulse described above. This curve is obtained by changing the phase $\phi = \phi_{\boldsymbol{p}}$ from 0 to 2π. Here, the blue line relates to the ramp-up portion of the pulse ($0 \leqslant \phi < \pi$) and the cyan line corresponds to its ramp-down portion ($\pi \leqslant \phi \leqslant 2\pi$). As it can be seen, the momentum spiral has several self-intersections which, according to our previous analysis, determine the phases for which interference-dominated patterns can be observed. In the right panel of the same figure, we present the one-dimensional spin-independent probability distribution $\mathcal{P}(\boldsymbol{p})$ [Eq. (10)] calculated along the path followed by the spiral of life, i.e., we show the function $\mathcal{P}(\phi) \equiv \mathcal{P}\big(\boldsymbol{p}(\phi)\big)$ for $\phi \in [0, 2\pi]$. This distribution acquires very small values at $\phi < 0.3\pi$ and $\phi > 1.7\pi$, i.e., when the laser field strength is small. Furthermore, in the region from $\phi \approx 0.6\pi$ to 1.4π, $\mathcal{P}(\phi)$ presents long plateaux between well-defined interference-dominated structures. The latter are located at phases which coincide with the self-intersections of the spiral. (Note that, in a coherent process, constructive interference would lead to probability distributions with peaks which are four times larger than the interference-free surroundings, which is actually the case.) Also, a soft valley in $\mathcal{P}(\phi)$, which occurs when the field strength is maximum, is evidenced. Such behavior could be a signature of the so-called stabilization against ionization (see, e.g., Ref. [16, 17, 18, 19]), as the valley becomes deeper with increasing the laser field intensity.

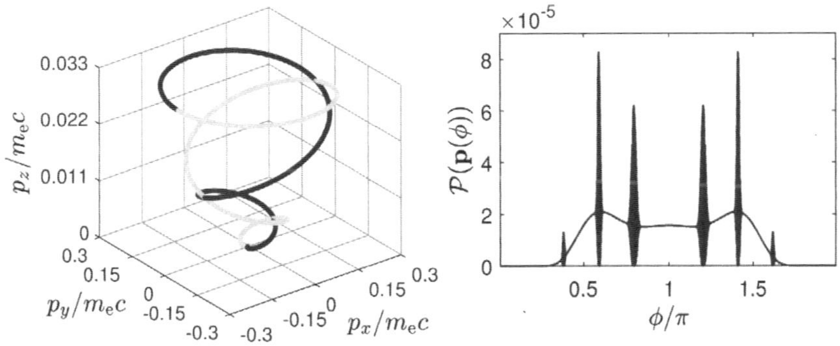

Figure 1. Three-dimensional spiral of life $\boldsymbol{p}(\phi)$ [Eq. (16)] (left panel) and the one-dimensional spin-independent probability distribution along such path, $\mathcal{P}(\boldsymbol{p}(\phi))$ [see, Eq. (10)] (right panel). The laser pulse is described in the text and the parameters used here are $I = 10^{17}$ W/cm^2, $N_{\rm osc} = 5$, and $\omega_{\rm L} = 1.5498\,$eV. Both curves are obtained by changing the phase $\phi = \phi_{\boldsymbol{p}}$ from 0 to 2π. For visual purposes, in the left panel the blue line relates to the ramp-up portion of the pulse and the cyan line to its ramp-down portion.

Now, in order to understand better the general features of the $\mathcal{P}(\phi)$ curve, we analyze the saddle points along the spiral. To this end, we calculate such points from Eq. (9) as a function of the phase, i.e., we are interested in the quantity $\phi_s(\phi)$. In the upper-left panel of Fig. 2 we show the imaginary part of the saddle point which contributes the most to the probability amplitude. It can be seen that $\mathrm{Im}\,\phi_s(\phi)$ is small at the middle of the pulse (i.e., in the region where the laser field strength is high). In the upper-right panel, we show the difference between the real part of the saddle point ($\mathrm{Re}\,\phi_s$) and $\phi_{\boldsymbol{p}} = \phi$. Such difference is negligible in the mid-portion of the pulse but its absolute value increases at the beginning and at the end of the pulse. Hence, we conclude that $\phi_s \approx \phi_{\boldsymbol{p}}$ is a good approximation provided that the phase ϕ is not close to zero or to 2π, i.e., when the strength of the pulse is large enough. In order to quantify this statement, in the lower-right panel of the same figure we present the time-dependent ponderomotive energy $U_p(\phi)$ [Eq. (12)] divided by the ionization potential, $I_p = m_{\rm e}c^2 - E_0$. Even though $U_p(\phi)$ acquires very small values at the beginning and at the end of the pulse, it increases rapidly and can be larger than $10^2 I_p$ at $\phi = \pi$. By comparing this curve with the upper panels, we conclude that our approximation $\phi_s \approx \phi_{\boldsymbol{p}}$ holds (i.e., $\mathrm{Im}\,\phi_s \approx 0$ and $\mathrm{Re}\,\phi_s \approx \phi_{\boldsymbol{p}}$) when $U_p(\phi) > 10 I_p$. Therefore, assuming that Eq. (18) is applicable in high-energy ionization, it can be said that the predictions arising from the three-dimensional spiral of life are valid provided that the photoelectron kinetic energies are larger than ten times the ionization potential (for the He^+ ion this value is around 0.54 keV), given the laser field parameters considered here. Such condition agrees with the range of applicability of the RSFA [see, Eq. (2)]. Finally, in the lower-left panel of Fig. 2 we present the imaginary part of the function $G(\phi_s)$ for the two most contributing saddle points. Note that, while the blue curve decreases in the middle part of the interval, the red one is considerably larger, except at well defined phases. Their points of contact coincide with the self-intersections of the three-dimensional spiral and appear at the exact same phases where the interference-dominated structures are located (see, the right panel of Fig. 1). Now, by comparing the behavior of $\mathrm{Im}\,G(\phi_s)$ (blue curve) and the one-dimensional probability distribution shown in Fig. 1, it can be seen that the phase intervals where the former increases coincide with the regions where $\mathcal{P}(\phi)$ is small. This is in agreement with the analysis which follows from the saddle-point treatment of Eq. (5).

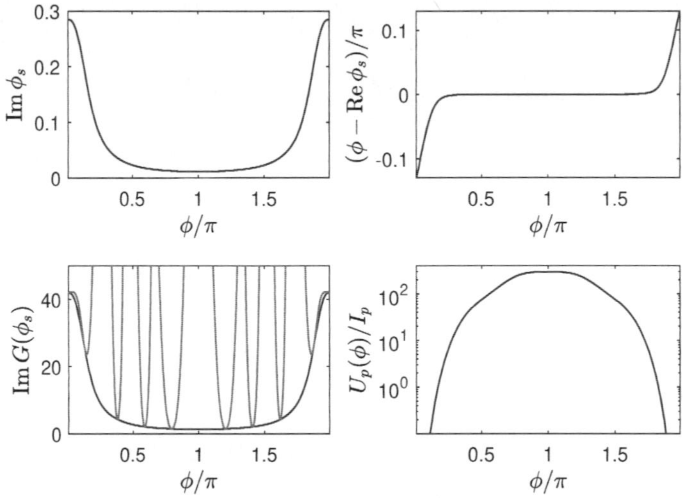

Figure 2. Behavior of the most relevant saddle points contributing to the probability amplitude of ionization along the three-dimensional spiral of life $\boldsymbol{p}(\phi)$ [Eq. (16)]. The laser field parameters are the same as in Fig. 1. In the upper panels we present the imaginary part of ϕ_s as a function of ϕ (left) and the difference $\phi - \mathrm{Re}\,\phi_s$ (right). The time-dependent ponderomotive curve $U_p(\phi)$ [Eq. (12)] in terms of the ionization potential, $I_p = m_e c^2 - E_0$, is also shown (lower-right panel). Finally, the function $\mathrm{Im}\,G(\phi_s)$ [see, Eq. (6)] evaluated at the two most contributing saddle points is presented (lower-left panel).

4. Spiral of life in classical electrodynamics

It is the aim of this Section to study the dynamics of photoelectrons in the laser field from the perspective of a purely classical theory. As we will show, the Newton-Lorentz equations determine a series of *invariant quantities of motion* which, in turn, lead to similar expressions as the ones obtained from the approximated saddle-point analysis (Sec. 2.1). Even though the results presented here are well-known (see, for instance, Refs. [20, 21, 22]), we will reformulate them in the spirit of this paper.

Let us start by considering a free electron traveling with velocity $\boldsymbol{v}$ and momentum $\boldsymbol{p}$ in the presence of a laser pulse. The latter is characterized by the oscillating electric and magnetic fields $\boldsymbol{\mathcal{E}}(\boldsymbol{x},t)$ and $\boldsymbol{\mathcal{B}}(\boldsymbol{x},t)$, respectively, and it propagates along the direction $\boldsymbol{n}$. For transverse waves, such as light, the relation $\boldsymbol{n}\cdot\boldsymbol{\mathcal{E}}(\boldsymbol{x},t) = \boldsymbol{n}\cdot\boldsymbol{\mathcal{B}}(\boldsymbol{x},t) = 0$ is always fulfilled. The classical equation of motion, which governs the dynamics of the electron, is

$$\frac{\mathrm{d}\boldsymbol{p}}{\mathrm{d}t} = e\boldsymbol{\mathcal{E}}(\boldsymbol{x},t) + e\boldsymbol{v}\times\boldsymbol{\mathcal{B}}(\boldsymbol{x},t). \tag{20}$$

Now, we assume that the electromagnetic field has the plane wave front, i.e., it depends on the phase, $\phi = \omega\,(t - \boldsymbol{n}\cdot\boldsymbol{x}/c)$. In what follows, we also assume that the field vanishes for $\phi \leqslant 0$ and $\phi \geqslant 2\pi$. We choose a gauge in which the vector potential $\boldsymbol{A}$ depends only on ϕ. Namely, we define it as

$$\boldsymbol{A}(\phi) = -\frac{1}{\omega}\int_0^{\phi} \mathrm{d}\phi'\,\boldsymbol{\mathcal{E}}(\phi'). \tag{21}$$

Note that this quantity fulfills the conditions $\boldsymbol{A}(0) = \boldsymbol{A}(2\pi) = \boldsymbol{0}$. Furthermore, as the electric and magnetic fields obey the relation $\boldsymbol{\mathcal{B}}(\phi) = \boldsymbol{n}\times\boldsymbol{\mathcal{E}}(\phi)/c$, or, $\boldsymbol{\mathcal{E}}(\phi) = c\boldsymbol{\mathcal{B}}(\phi)\times\boldsymbol{n}$, the vector potential can also be written in terms of $\boldsymbol{\mathcal{B}}(\phi)$,

$$\boldsymbol{A}(\phi) = -\frac{c}{\omega}\int_0^{\phi} \mathrm{d}\phi' \, \boldsymbol{\mathcal{B}}(\phi') \times \boldsymbol{n}. \tag{22}$$

In the following, we assume that the particle four-momentum is on the effective mass shell, i.e., $p \cdot p = (p^0)^2 - \boldsymbol{p}^2 = (m_{\mathrm{eff}}c)^2$. This implies that the velocity and momentum of the electron are related according to $\boldsymbol{v} = c\boldsymbol{p}/p^0$. With this in mind, we rewrite the magnetic part of the Lorentz force exerted by the laser field as

$$\boldsymbol{v} \times \boldsymbol{\mathcal{B}}(\phi) = \frac{1}{p^0}\boldsymbol{p} \times [\boldsymbol{n} \times \boldsymbol{\mathcal{E}}(\phi)] \equiv \frac{1}{p^0}\Big([\boldsymbol{p} \cdot \boldsymbol{\mathcal{E}}(\phi)]\, \boldsymbol{n} - (\boldsymbol{p} \cdot \boldsymbol{n})\, \boldsymbol{\mathcal{E}}(\phi)\Big). \tag{23}$$

Hence, the Newton-Lorentz equation of motion takes the form

$$\frac{\mathrm{d}\boldsymbol{p}}{\mathrm{d}\phi} \equiv \frac{\mathrm{d}\boldsymbol{p}^{\perp}}{\mathrm{d}\phi} + \frac{\mathrm{d}p^{\parallel}}{\mathrm{d}\phi}\boldsymbol{n} = \frac{1}{\omega}e\boldsymbol{\mathcal{E}}(\phi) + \frac{1}{\omega(p^0 - p^{\parallel})}\left[e\boldsymbol{\mathcal{E}}(\phi) \cdot \boldsymbol{p}^{\perp}\right]\boldsymbol{n}, \tag{24}$$

where ϕ is considered the independent variable. This is the starting point of our further analysis.

4.1. Invariants of motion

As presented above, the Newton-Lorentz equations form a system of three first-order differential equations and, therefore, there should be at most three independent invariants of motion. In this Section, we will analyze the properties of (24), which lead to the determination of those invariants. Let us start by examining the parallel component of the momentum, $p^{\parallel} = \boldsymbol{p} \cdot \boldsymbol{n}$. It follows directly from Eq. (24) that

$$\frac{\mathrm{d}p^{\parallel}}{\mathrm{d}\phi} = \frac{\mathrm{d}\boldsymbol{p}}{\mathrm{d}\phi} \cdot \boldsymbol{n} = \frac{e\boldsymbol{\mathcal{E}}(\phi) \cdot \boldsymbol{p}^{\perp}}{\omega(p^0 - p^{\parallel})}. \tag{25}$$

Furthermore, as the derivative of p^0 with respect to the phase ϕ is given by

$$\frac{\mathrm{d}p^0}{\mathrm{d}\phi} = \frac{\boldsymbol{p}}{p^0} \cdot \frac{\mathrm{d}\boldsymbol{p}}{\mathrm{d}\phi} = \frac{e\boldsymbol{\mathcal{E}}(\phi) \cdot \boldsymbol{p}^{\perp}}{\omega(p^0 - p^{\parallel})} \tag{26}$$

and, according to Eq. (25), we obtain

$$\frac{\mathrm{d}}{\mathrm{d}\phi}(p^0 - p^{\parallel}) = 0. \tag{27}$$

This leads to the conclusion that the quantity $p^0 - p^{\parallel}$ does not change during the motion of the electron in the laser field. Therefore, the first invariant, denoted as I_1, is

$$I_1 = p^0 - p^{\parallel}. \tag{28}$$

Let us now go back to Eq. (24). From there, it is evident that the first derivative of the perpendicular component of momentum with respect to ϕ is given by

$$\frac{\mathrm{d}\boldsymbol{p}^{\perp}}{\mathrm{d}\phi} = \frac{e\boldsymbol{\mathcal{E}}(\phi)}{\omega}, \tag{29}$$

which, in terms of the vector potential, reads

$$\frac{\mathrm{d}}{\mathrm{d}\phi}[\boldsymbol{p}^{\perp} + e\boldsymbol{A}(\phi)] = \boldsymbol{0}. \tag{30}$$

Thus, the second invariant of motion is defined as

$$\boldsymbol{I}_2 = \boldsymbol{p}^{\perp} + e\boldsymbol{A}(\phi) = \boldsymbol{p}^{\perp} - \frac{e}{\omega}\int_0^{\phi} \mathrm{d}\phi' \boldsymbol{\mathcal{E}}(\phi'). \tag{31}$$

This quantity is the perpendicular part of the canonical momentum $\boldsymbol{p} + e\boldsymbol{A}(\phi)$.

Finally, the next invariant is obtained from the fact that the relation

$$\frac{\mathrm{d}}{\mathrm{d}\phi}\left[\frac{1}{2}(\boldsymbol{p}^{\perp})^2 - p^{\parallel}(p^0 - p^{\parallel})\right] = 0 \tag{32}$$

always holds. Therefore, we define I_3 as

$$I_3 = p^{\parallel}(p^0 - p^{\parallel}) - \frac{1}{2}(\boldsymbol{p}^{\perp})^2. \tag{33}$$

Note, however, that $\boldsymbol{I}_2$ represents two independent quantities and, as a consequence, some of the invariants should be related. This is indeed the case, as one can check that $I_1^2 + 2I_3 = m_{\mathrm{eff}}^2 c^2$. Hence, I_1 and I_3 define the effective mass of the electron.

4.2. Physical consequences arising from the invariants of motion

The invariants of motion defined in the previous Section can be used to understand the dynamics of a charged particle in the laser field, at least from a classical perspective. For instance, starting from Eq. (31) and remembering that the vector potential vanishes at the beginning and at the end of the pulse, we can conclude that the initial and final perpendicular components of the momentum of a free electron are the same, i.e., $\boldsymbol{p}^{\perp}(\phi = 0) = \boldsymbol{p}^{\perp}(\phi = 2\pi)$. Furthermore, since $p^0 - p^{\parallel}$ is another invariant of motion, we also obtain $p^{\parallel}(\phi = 0) = p^{\parallel}(\phi = 2\pi)$. This actually means that a free electron will not be accelerated or decelerated by the plane-wave-fronted pulse, as its initial and final four-momenta are the same. Such observation is in agreement with the Lawson-Woodward (no acceleration) theorem [23, 24].

Now, let us assume that for some phase $\phi_s \in]0, 2\pi[$ an electron is created with initial momentum $\boldsymbol{p}(\phi_s) = \boldsymbol{0}$. Then, after the pulse is over ($\phi = 2\pi$), the invariant $\boldsymbol{I}_2$ implies that

$$\boldsymbol{p}^{\perp}(\phi = 2\pi) = e\boldsymbol{A}(\phi_s). \tag{34}$$

On the other hand, I_1 and I_3 determine $p^{\parallel}$,

$$p^{\parallel}(\phi = 2\pi) = \frac{(\boldsymbol{p}^{\perp})^2}{2m_{\mathrm{eff}}c} = \frac{e^2\boldsymbol{A}^2(\phi_s)}{2m_{\mathrm{eff}}c}. \tag{35}$$

Moreover, at the end of the pulse, the zeroth-component of the four-momentum is given by

$$p^0(\phi = 2\pi) = m_{\mathrm{eff}}c + p^{\parallel}(\phi = 2\pi) = m_{\mathrm{eff}}c + \frac{e^2\boldsymbol{A}^2(\phi_s)}{2m_{\mathrm{eff}}c}. \tag{36}$$

As we are considering a purely classical treatment, we have that $m_{\mathrm{eff}} = m_{\mathrm{e}}$.

Let us now compare the classical results obtained in this Section [Eqs. (34), (35), and (36)] with the set of equations derived from the analytical approximation of the saddle points [Eqs. (14), (15), and (18)]. In doing so, it is important to remember that the real part of the saddle points are related to the time of birth of the electron in the continuum, i.e., those are the phases for which an electron is created with zero momentum. First, one can see that Eq. (14) is the exact same as its classical counterpart (34), with the 'phase of birth' ϕ_s replaced by the approximated saddle point $\phi_{\boldsymbol{p}}$. Moreover, Eqs. (15) and (18) differ from (35) and (36) only by the definition of the effective mass (while in the classical treatment $m_{\mathrm{eff}} = m_{\mathrm{e}}$, in the saddle-point approximation $m_{\mathrm{eff}} = q^0/c \equiv E_0/c^2$). We, therefore, conclude that the three-dimensional spiral of life presented in Sec. 2.1 is recovered from classical considerations provided that $q^0/c \approx m_{\mathrm{e}}$, i.e., for light ions.

5. Conclusions

We have shown that the three-dimensional spiral of life, which arises from an approximate solution to the saddle-point equation, is a very powerful tool in understanding the high-energy photoionization. It determines the possible electron momenta for which the probability of ionization is non-negligible. Also, the shape of the spiral can be used to predict the formation (or absence) of interference-dominated structures by locating its self-intersections. Finally, we demonstrated that this three-dimensional curve is also obtained from a purely classical theory, provided that the ions are light.

Acknowledgments

This work is supported by the National Science Centre (Poland) under Grant No. 2014/15/B/ST2/02203.

References

[1] https://eli-laser.eu
[2] www.xcels.iapras.ru
[3] Reiss H R 1990 *Phys. Rev.* A **42** 1476
[4] Reiss H R 1990 *J. Opt. Soc. Am.* B **7** 574
[5] Krajewska K and Kamiński J Z 2016 *Phys. Rev.* A **94** 013402
[6] Cajiao Vélez F, Kamiński J Z and Krajewska K 2017 *J. Phys. B: At. Mol. Opt. Phys.* https://doi.org/10.1088/1361-6455/aaa056
[7] Cajiao Vélez F, Kamiński J Z and Krajewska K 2017 *J. Phys.: Conf. Ser.* **826** 012010
[8] Krajewska K, Cajiao Vélez F and Kamiński J Z 2017 *Proc. of SPIE* **10241** 102411J
[9] Kamiński J Z, Cajiao Vélez F and Krajewska K 2017 *Laser Phys. Lett.* **14** 075301
[10] Krajewska K, Cajiao Vélez F and Kamiński J Z 2017 *EPL* **119** 13001
[11] Wolkow D M 1935 *Z. Phys.* **94** 250
[12] Krajewska K and Kamiński J Z 2012 *Phys. Rev.* A **85** 062102
[13] Krajewska K and Kamiński J Z 2012 *Phys. Rev.* A **86** 052104
[14] Bjorken J D and Drell S D 1964 *Relativistic Quantum Mechanics* (New York: McGraw-Hill)
[15] Krajewska K and Kamiński J Z 2015 *Phys. Rev.* A **92** 043419
[16] Gavrila M and Kamiński J Z 1984 *Phys. Rev. Lett.* **52** 613
[17] Fedorov M V and Movsesian A M 1988 *J. Phys.* B **21** L155
[18] Popov A M, Tikhonova O V and Volkova E A 2003 *J. Phys.* B **36** R125
[19] Bogatskaya A V and Popov A M 2015 *Laser Phys. Lett.* **12** 045303
[20] Landau L D and Lifshitz E M 1975 *The Classical Theory of Field* (New York: Pergamon)
[21] Ortner J and Rylyuk V M 2000 *Phys. Rev.* A **61** 033403
[22] Wooten R E and Macek J H 2004 *Am. J. Phys.* **72** 998
[23] Woodward P M 1946 *J. Inst. Electr. Eng., Part IIIA* **93** 1554
[24] Lawson J D 1979 *IEEE Trans. Nucl. Sci.* **26** 4217

The Multidisk Diode-Pumped High Power Yb:YAG Laser Amplifier of High-Intensity Laser System with 1 kHz Repetition Rate

G V Kuptsov[1,2], V V Petrov[1,2,3], V A Petrov[1,3], A V Laptev[1], A V Kirpichnikov[1], E V Pestryakov[1]

[1]Institute of Laser Physics SB RAS, 15b Ave. Ac. Lavrentyeva, Novosibirsk, 630090, Russia;
[2]Novosibirsk State National Research University, 2 St. Pirogova, Novosibirsk, 630090, Russia;
[3]Novosibirsk State Technical University, 20 Ave. K. Marksa, Novosibirsk, 630073, Russia

E-mail: kuptsov.gleb@gmail.com

Abstract. The source of instabilities in the multidisk diode-pumped high power Yb:YAG laser amplifier with cryogenic closed-loop cooling in the laser amplification channel of the high-intensity laser system with 1 kHz repetition rate was determined. Dissected copper mounts were designed and used to suppress instabilities and to achieve repeatability of the system. The equilibrium temperature dependency of the active elements on average power was measured. The seed laser for the multidisk amplifier was numerically simulated and designed to allow one to increase pulses output energy after the amplifier up to 500 mJ.

1. Introduction

In recent years, high power laser systems and their applications made a prominent progress. In spite of such laser systems are broadly used, at the present time, there are many studies targeted at the development and advancement of simultaneously high average and high peak power laser systems. Remarkable attention is attracted owing to the fact that fundamental and applied science require simultaneously high average and high peak power laser radiation, for instance, to generate high flux attosecond pulses [1], to create novel and effective sources of charged particle beams, to produce ultra-bright X and γ-rays [2] and high XUV flux [3] and other applications. Besides the development the principles and methods of creation, it is highly important to increase the laser system effectiveness [4]. To solve this problem, one can use modern high efficient diode laser sources as a pump and cryogenic cooling of active elements. Moreover, cryogenic cooling allows one to obtain high average and high peak power radiation while keeping beam quality [5]. In addition, there are extensive researches aimed at discovery and development of new optical media, laser and nonlinear as well, to fulfill continuous demands in the field of high-power laser systems [6].

Presently, at the Institute of Laser Physics of the Siberian Branch of the RAS an all diode-pumped all solid state high-intensity laser system with 1 kHz repetition rate is developed. The layout of the system is shown in Fig. 1.

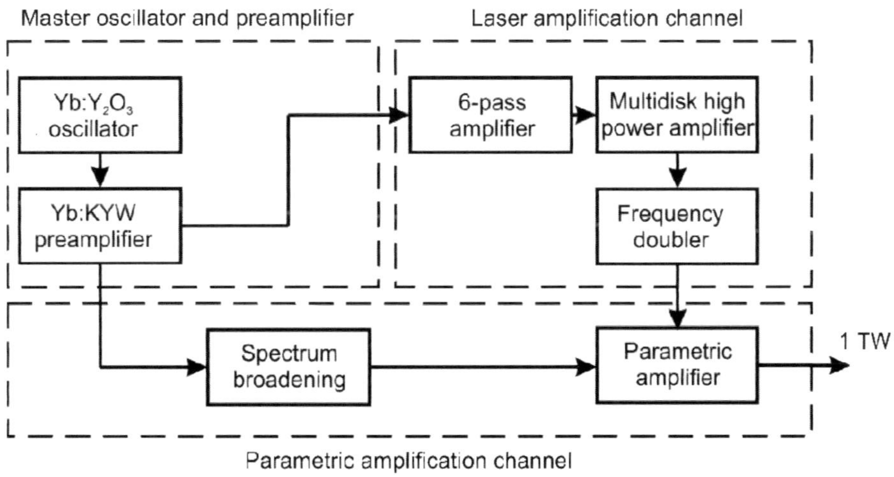

Figure 1. The layout of the high-intensity laser system with 1 kHz repetition rate

The first stage of our system consists of master oscillator, stretcher and regenerative preamplifier. The master oscillator is based on Yb:Y_2O_3 ceramic cooled with liquid nitrogen. It produces ~1 nJ, ~1 ps pulses with spectral width of ~1.7 nm FWHM (full width at half maximum) centered at 1030 nm with ~80 MHz repetition rate. In fibre stretcher (not shown in Fig. 1) the pulses are at the same time spectrally widened to 6 nm FWHM and temporally stretched to 500 ps [7]. These pulses are picked and amplified at 1 kHz repetition rate in the Yb:KYW regenerative preamplifier to 0.5 mJ. After the regenerative preamplifier, the beam is split to two optically synchronized parallel channels: laser amplification channel and parametric amplification channel.

The parametric amplification channel has highly nonlinear photonic crystal fibre[8] and optical parametric amplifier [9]. The input signal is broadened to ~150 nm FWHM in photonic-crystal fibre what allows one to obtain ~10 fs pulses after compression. After broadening, the pulses are to be amplified to 10 mJ, which corresponds to 1 TW of peak power at the output of the system. The parametric amplification channel takes the radiation from the laser amplification channel as the pump. This method keeps synchronization between pump and seed pulses effortlessly, thus, provides reliable stability while retaining high output intensity and high repetition rate [10].

Two serial all-diode pumped multipass laser amplifiers and a frequency doubling unit form the laser amplification channel. It is elaborated to provide150 mJ pulses with central wavelength 515 nm and 1 kHz repetition rate. The first amplifier is water cooled 6-pass Yb:YAG amplifier which takes a 0.5 mJ pulse and amplifies it to 10 mJ. The second amplifier, multidisk diode-pumped high power Yb:YAG laser amplifier, is closed-loop cryogenically cooled and designed to boost pulses energy up to 300 mJ.

2. The cooling system of the multidisk amplifier

In this paper, the multidisk diode-pumped high power Yb:YAG laser amplifier cooling system is discussed. The amplifier has eight diffusion-bonded Yb:YAG (10 at.% Yb) crystals mounted in pairs to four independent massive copper heatsinks.

Copper heatsinks are attached to cryogenic pulse tube refrigerators. The light travels more than 30 m inside the amplifier and undergoes 32 passes through the active elements. This leads to high sensitivity to angular displacements and drifts of the active elements. Closed-loop cryogenic cooling based on pulse tubes provides stable temperature of the crystals and robust operation, however, at the output of the amplifier we observed angular cyclic displacements of beam centroid with magnitude of ~40 urads. The amplifier self-misaligned itself after each start of cooling. Either displacements and self-misaligning are proven to be caused by pulse tubes, despite the fact that the pulse tubes have extremely low vibrations in the operation region.

The maximal average pump power used in the experiments for a single heatsink with two active elements attached is 200 W (2x100 W). The heat load therefore is 0-20 W since the quantum defect of Yb ion is ~0.1. The pump pulses have next parameters: power is 0-200 W (peak), repetition rate is 0.1-1 kHz, central wavelength is 936 nm, duration is 500 us long and the temporal pulse shape is rectangular. The minimal achievable temperature in the absence of pump is 40 K. The equilibrium temperature of the non-dissected, solid copper heatsink while pumped with maximum pump power is 120 K. According to simple estimation based on Boltzmann factor for Yb:YAG, one should not let active elements be hotter than ~130 K to preserve amplification efficiency. Also, it is obviously important to remove heat fast in the presence of pump to prevent breakdown of crystals. Based on calculations [11], dissected copper heatsinks were proposed to supress the vibrations while preserving thermal conductivity. The dissected heatsink has two parts: a copper plate matching cryocooler operation region and a massive copper cube. The cube is host to active elements, mounted on the opposite sides of the cube. 12 brazed copper braids connect cube and plate. Each copper braid has cross-section area of 10 mm^2.

Thermal properties of the dissected heatsinks were determined. The equilibrium temperature was tested under different pump regimes. With dissected heatsinks, the equilibrium temperature of 120 K is reached when the average pump power equals 100 W (maximal peak pump power but at 500 Hz instead of 1 kHz). This means that the experimental results are not in agreement with calculations approximately by a factor of two, however there is good qualitative agreement. It is supposed that the cause is manufacturing process, notably brazing of contacts.

We have also proven experimentally that the dissected heatsinks have eliminated the self-misaligning problem and reduced cyclic displacements by a factor of 3 down to ~10 urad. The results of measuring the displacements before and after heatsinks modification are given in Fig. 2.

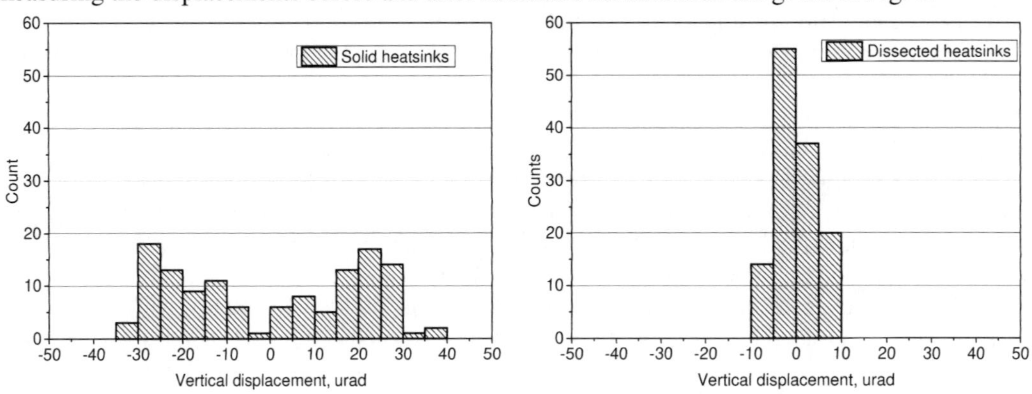

Figure 2. The histogram of the vertical angular cyclic displacements of beam centroid: a - using solid heatsinks, b - using dissected heatsinks.

3. The alternative seed laser of the multidisk amplifier

The estimation shows that one can store more than 500 mJ in the 8 active elements using our diode pumping at 1 kHz. To increase efficiency and simplify the setup, an alternative approach is proposed to amplify the pulse to ~500 mJ instead of 300 mJ at the output of the multidisk amplifier. Therefore, the efficiency of extraction will be almost a unity. Calculations show that a seed laser instead of the first multipass amplifier allows achieving 500 mJ at the output. So, one need a laser that produces pulses having next parameters: energy >30 mJ, duration 2-10 ns, repetition rate 1 kHz and central wavelength 1030 nm.

To determine the best laser design, laser rate equations (1) were solved numerically using Runge-Kutta-Fehlberg algorithm for different laser configurations [12,13]. The laser rate equations system is as follows:

$$\begin{cases} \dfrac{d\phi}{dt} = \left(V_g B_g N_g - V_s B_s N_s - V_s (1-N_s) B_{se} - \dfrac{1}{\tau_c} - Q(t) \right) \phi \\ \dfrac{dN_g}{dt} = W_p(t) \cdot (1-N_g) - N_g \left(B_g \phi + \dfrac{1}{\tau_g} \right) \\ \dfrac{dN_s}{dt} = \dfrac{1-N_s}{\tau_s} - B_s N_s \phi \end{cases} \tag{1}$$

where ϕ – photon flux; N – population inversion, m^{-3}; W_p – pump rate, s^{-1}; B – emission rate, s^{-1}; V – mode volume, m^3; τ_c – photon cavity lifetime, s; τ – fluorescence lifetime, s; Q(t) – active losses (i.e. gate) switching function, s^{-1}; g, s and se indices stand for gain, saturable absorber and saturable absorber excited state, respectively.

All temperature-dependent parameters were taken into account in the modeling. Also, resonator roundtrip losses were considered 15 %, gate reflectivity – 4 %, gate rise/fall times – 5 ns. Considering the numerical data obtained, an actively Q-switched Yb:YAG laser with passive mode-locking in a Cr^{4+}:YAG saturable absorber was chosen. An electro-optical shutter defines the repetition rate and the saturable absorber forms the pulse duration and shape. Moreover, the laser pulse is not chirped but its duration is few nanoseconds, thus is suitable for parametric amplification pumping, and there is no need to use stretcher-compressor system. Furthermore, it does not collapse in time domain during laser amplification due to spectral narrowing, the well known issue in the cryogenic laser amplifiers.

The best proposed design supposes usage one of the disks of the second multipass laser amplifier as the active element of the alternative seed laser. The huge advantage of this concept is perfect matching of the spectra of the seed laser and the amplifier since their active elements are at the same temperature. The spectral match increases efficiency, eliminates spectral distortion and prevents spectral shape distortions of the amplified pulse. The results of the modeling are shown in Fig.3.

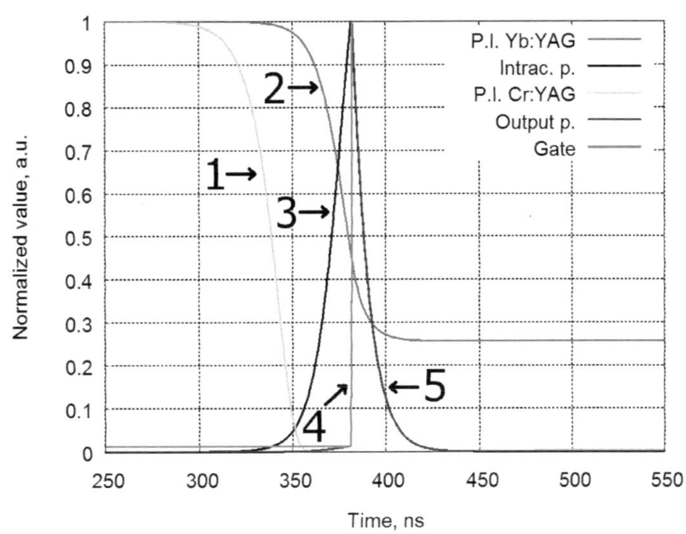

Figure 3 (in colour online). The results of modelling of intracavity dynamics.
Curves stand for: 1 (green) - the population of inversion (P.I.) of the saturable absorber; 2 (red) - the population of inversion of the active medium; 3 (black) - the intracavity power; 4 (magenta) - the gate losses; 5 (blue) - the output power.

The optimal parameters of the laser, pursuant to modeling, are: cavity length is 3 m, beam diameter at the active element is 1.25 mm, gate pulse duration is 380 ns, the initial transmission of the absorber is 70% and its thickness is 5 mm. The absorber is also cooled with the same closed-loop system, being mounted to the same dissected copper heatsink. The estimated pulse energy is 34 mJ and pulse duration is ~8 ns.

4. Results

The source of mechanical instabilities in the multidisk diode-pumped high power Yb:YAG laser amplifier with cryogenic closed-loop cooling was determined. The experimental and simulation data were used to design dissected copper mounts to suppress instabilities of cryogenic closed-loop cooling. The angular displacements are suppressed by a factor of 3, making the radiation at the output of the multidisk amplifier acceptable for pumping a parametric amplifier. Moreover, the repeatability of the system between room temperature-cryogenic temperature-room temperature cycles is achieved.

The equilibrium temperature dependency of the active elements on average power was measured. The system is able to operate with diode pump pulse repetition rate of 500 Hz. The data is used for further enhancement of the closed-loop cryocooling system to achieve 1 kHz repetition rate.

The seed laser for the multidisk amplifier with cryogenic closed-loop cooling was numerically simulated and designed. The optimal parameters of the designed laser are determined. Calculation shows that it is possible to achieve 500 mJ pulses at the output of the laser amplification channel if one use the alternative seed laser.

All the data, both simulated and experimental, are used in the development of the all diode-pumped high intensity cryogenic laser system working with 1 kHz repetition rate.

Acknowledgements

This work is supported in part by RAS Program "Extreme laser radiation: physics and fundamental applications", grant 0307-2015-0018 and Government program, 0307-2016-0006.

References

[1] Emaury F, Diebold A, Saraceno C J and Keller U 2015 *Optica* **2** 980-984

[2] Chvykov V, Cao H, Nagymihaly R, Kalashnikov M P, Khodakovskiy N, Glassock N, Ehrentraut L, Schnuerer M and Osvay K. 2016 *Opt. Lett.* **41** 3017-3020

[3] Rothhardt J, Hädrich S, Seise E, Krebs M, Tavella F, Willner A, Düsterer S, Schlarb H, Feldhaus J, Limpert J, Rossbach J and Tünnermann A 2010 *Optics Express* **18** 12719

[4] Mourou G A, Hulin D and Galvanauskas A 2006 *AIP Conference Proceedings Third International Conference on Superstrong Fields in Plasmas* **827** 152-163

[5] Brown D C, Tornegard S and Kolis J 2016 *High Power Laser Science and Engineering* **4** e15

[6] Bagayev S N, Osipov V V, Pestryakov E V, Solomonov V I, Shitov V A, Maksimov R N, Orlov A N, Petrov V V 2015 *J. Appl. Mechanics and Tech. Physics* **56** 150-157

[7] Kuptsov G V, Petrov V V, Laptev A V, Petrov V A and Pestryakov E V 2016 *Quantum Electronics* **46** 801-805

[8] Kuptsov G V, Petrov V A, Laptev A V,, Petrov V V and Pestryakov E V 2014 *Laser Optics 2014: 16 intern. conf.* 305

[9] Kuptsov G V, Petrov V V, Petrov V A, Kirpichnikov A V, Laptev A V and Pestryakov E V 2016 *Журнал прикладной спектроскопии* **83** 533-534

[10] Petrov V V, Pestryakov E V, Laptev A V, Petrov V A, Kuptsov G V, Trunov V I and Frolov S A 2014 *Quantum Electronics* **44** 452–457

[11] Petrov V A, Kuptsov G V, Petrov V V, Kirpichnikov A V, Laptev A V and Pestryakov E V 2017 *AIP Conference Proceedings* **1893** 030121

[12] Qin-Yong Z, Yong W, Ji-Chuan X and Da-Yong Z 2011 *Chin. Phys. B* **20** 034204

[13] Tsunekane M and Taira T 2013 *IEEE Journal of Quantum Electronics* **49** 454-461

IOP Conf. Series: Journal of Physics: Conf. Series **999** (2018) 012009 doi:10.1088/1742-6596/999/1/012009

Optimization of CW Fiber Lasers With Strong Nonlinear Cavity Dynamics

O.V. Shtyrina[1,2], S.A. Efremov[1,2], I.A. Yarutkina[1,2], A.S. Skidin[1,2], M.P. Fedoruk[1,2]

[1]Novosibirsk State University, 630090, 2 Pirogova str., Novosibirsk, Russia
[2]Institute of Computational Technologies SB RAS, 630090, 6 Lavrentiev ave., Novosibirsk, Russia

E-mail: efremov.math@gmail.com

Abstract. In present work the equation for the saturated gain is derived from one-level gain equations describing the energy evolution inside the laser cavity. It is shown how to derive the parameters of the mathematical model from the experimental results. The numerically-estimated energy and spectrum of the signal are in good agreement with the experiment. Also, the optimization of the output energy is performed for a given set of model parameters.

1. Introduction

In general, the optimization of the signal evolution inside the strong nonlinear laser cavity cannot be performed using the analytical approximation. To carry out such a multiparametric optimization, the numerical modeling based on the coupled modified Schrodinger equations is employed [1]. However, not all the model parameters can be derived from the experimental results. Because of the large number of parameters and the numerical modeling complexity, there is a need in developing the simplified models of the saturation gain.

To describe the spatial-temporal evolution of the electromagnetic field, the following coupled modified Schrodinger equations with the gain saturation are used:

$$\pm\frac{\partial A^{\pm}}{\partial z} = -i\beta_1\frac{\partial A^{\pm}}{\partial t} - i\frac{\beta_2}{2}\frac{\partial^2 A^{\pm}}{\partial t^2} + i\gamma\left(|A^{\pm}|^2 + 2|A^{\mp}|^2\right)A^{\pm} + \frac{g}{2\left(1 + P/P_{SAT}\right)}A^{\pm}. \quad (1)$$

Here the signs "$\pm$" denote the forward and backward field directions, respectively, β_1, β_2, γ are dispersion and nonlinear coefficients [1]. The power is denoted by $P = P^{+} + P^{-} = \frac{1}{T_R}\left(\int |A^{+}|^2\, dt + \int |A^{-}|^2\, dt\right)$, T_R is the round-trip time, g and P_{SAT} are the small signal gain and saturation power, respectively. The values of g and P_{SAT} are derived from the experimental measurements using the method described below. To numerically solve the system, the split-step Fourier method was used.

Besides the system of NLS equations, the one-level gain media model is used:

$$\frac{dP^{\pm}}{dz} = \pm\left(\frac{g}{1 + P/P_{SAT}}\right)P^{\pm}, \quad P(z) = P^{+}(z) + P^{-}(z). \quad (2)$$

IOP Conf. Series: Journal of Physics: Conf. Series **999** (2018) 012009 doi:10.1088/1742-6596/999/1/012009

The experimental setup is shown in figure 1 [1]; it represents itself the Yb-doped fiber laser of the Fabry-Perot type that consists of the high-reflectivity fiber Bragg grating (R^-) and the 4% back reflection (R^+) from the fiber edge at the left and the right sides respectively. The active fiber length is $L = 35$ m. From the left side of the cavity the pump is produced by the 23 W laser diode at 910 nm. The output signal power of such setup can be up to 12 W with spectrum width 0.5 nm and the maximum spectral power at 1084 nm. The experimental results are shown on figure 2.

Figure 2 shows the energy balance inside the laser cavity. Here one can notice that at the free end the output energy is much higher, than at the FBG: it means that FBG has much higher reflectivity. The FBGs refraction spectrum is shown on figure 3.

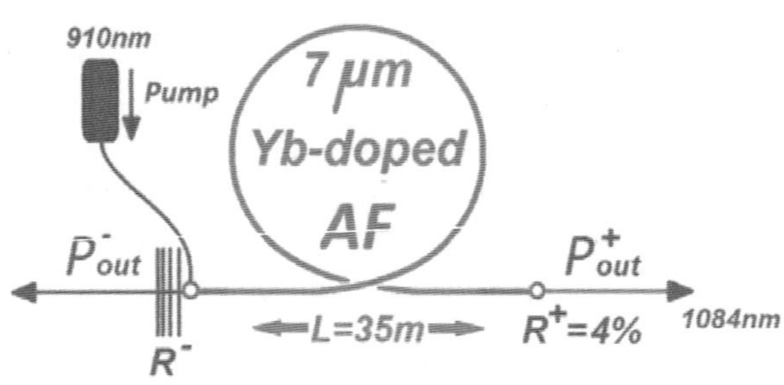

Figure 1. The experimental resonator scheme.

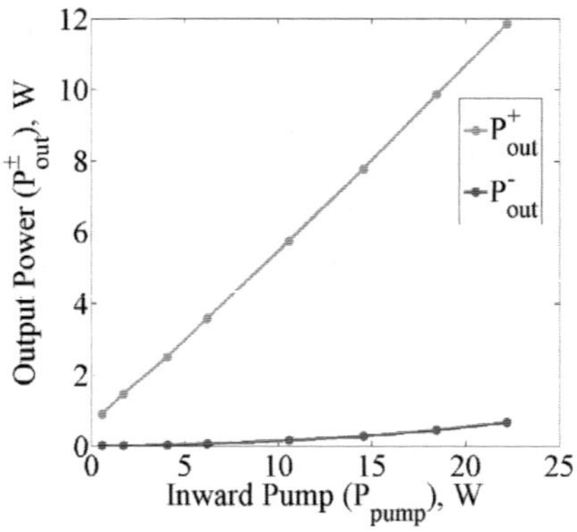

Figure 2. The experimental output signal power.

To perform the numerical simulation (particularly, to perform the optimization), it is necessary to know the parameters of the model. However, in many cases some parameters cannot be

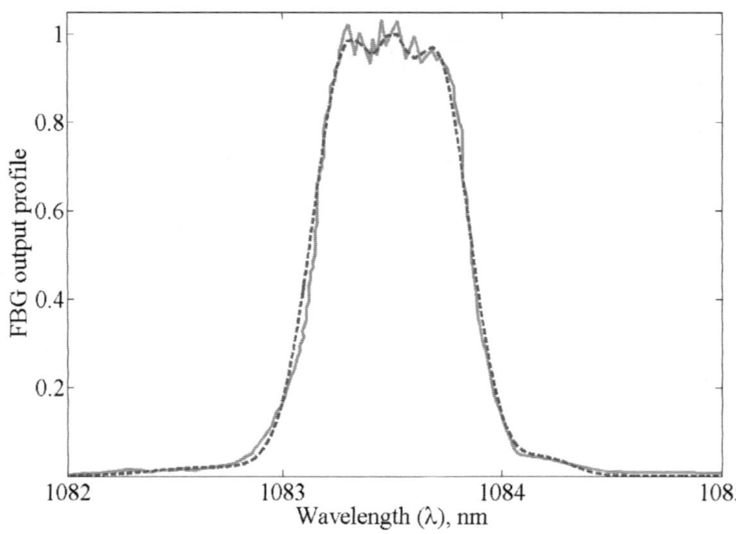

Figure 3. The FBGs refraction spectrum: measured (solid blue line) and approximated (dashed purple line).

derived directly from the experimental results. Such parameters can be obtained by means of the theoretical approximation.

2. Saturated gain equation

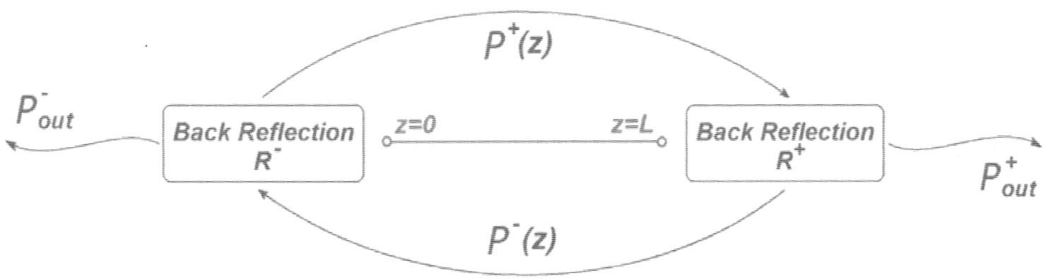

Figure 4. A simple scheme of the energy balance in the laser cavity.

Let us derive a signal gain coefficient. When considering the signal gain, it is normally believed that the gain of the signal propagating in forward and backward directions is the same ([1],[2]). Therefore, the signal gain can be expressed as follows:

$$G = \frac{P^+(L)}{P^+(0)} = \frac{P^-(0)}{P^-(L)}.$$

It should be noted that the experimental results do not normally contain the information about the signal gain. But in mathematical modeling and the energy balance analysis the gain of the radiation through the laser cavity is one of the most important characteristics. Below in this section we derive the equation of the experimental gain coefficient that takes into account the structure of the linear laser cavity.

In case of the linear cavity, there are two reflection coefficients at the left (R^-) and the right (R^+) sides, the radiation propagates in both directions; hence, the following equations take place:

$$P^-(L) = GR^+P^+(0), \quad P^+(0) = GR^-P^-(L).$$

These expressions indicate that to achieve a stable generation, the signal gain should compensate the intracavity losses. It means that

$$G = \frac{1}{\sqrt{R^+R^-}}. \tag{3}$$

Thus, the output power can be expressed as follows:

$$P^+_{out} = G\left(1 - R^+\right)P^+(0),$$
$$P^-_{out} = G\left(1 - R^-\right)P^-(L).$$

After some simple transformations one can obtain that

$$\frac{P^+(L)}{P^+(0)} = G = \frac{P^+_{out}}{P^-_{out}} \cdot \frac{R^+G^2 - 1}{1 - R^+},$$

and the following saturated gain equation can be obtained:

$$G^2 - \frac{P^-_{out}}{P^+_{out}}\left(\frac{1}{R^+} - 1\right)G - \frac{1}{R^+} = 0. \qquad (4)$$

A physically consistent (positive) solution always exists and is derived from the following expression:

$$G = \frac{1}{2}\left[\frac{P^-_{out}}{P^+_{out}}\left(\frac{1}{R^+} - 1\right) + \sqrt{\left[\frac{P^-_{out}}{P^+_{out}}\left(\frac{1}{R^+} - 1\right)\right]^2 + \frac{4}{R^+}}\right].$$

Here one can notice that in the final equation there is no left-end reflection (R^-); for this reason there is no need to estimate this coefficient.

On figure 5 the signal gain estimated using equation 4 is shown. It can be seen that the

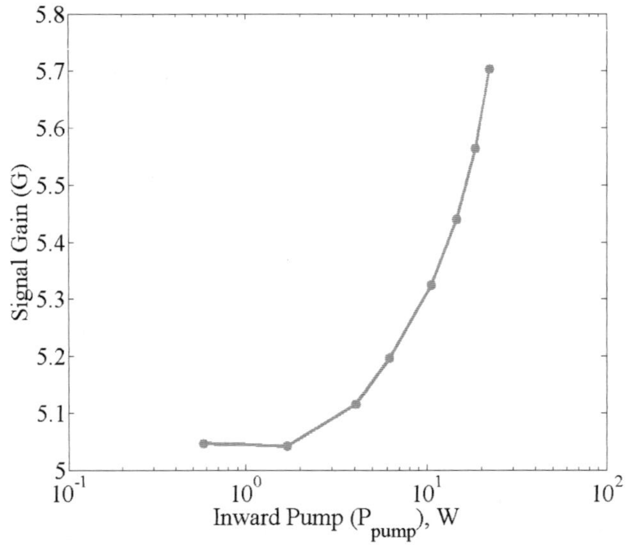

Figure 5. The signal gain computed from the experimental data.

gain slightly declines when the pump power is $P_{pump} = 1.7$ W. The reason for such behaviour may lie in a measurement error, this fact shows that although theory shown in previous section allows to exclude some parameters measured in experiment, equation 4 is still sensitive to the measurement accuracy.

The equation derived in this section enables to estimate the signal gain as a function of only the output signal power measured experimentally.

3. Saturated gain model

In present work the one-level gain model was chosen. Such model correctly describes the energy dynamics inside a laser cavity of chosen configuration. This model includes background losses, but here they can be neglected. Experimental results show, that generation area of such laser for a big range of inward pump (from 0.6 W up to 23 W) includes tiny variations of the signal gain (given in 4), and gain value corresponds to saturated regime. Therefore there is no need to take into consideration the background losses. Moreover in the experiment discussed here the active fiver length was always kept constant, so the small signal gain can be considered as

invariable. Here the small signal gain value was chosen as $gL = 10$ dB.
After reasoning given above one comes to the system of balance equations

$$\frac{dP^{\pm}}{dz} = \pm\left(\frac{g}{1+P/P_{SAT}}\right)P^{\pm}$$

which appears to be two differential equations with separable variables and can be easily integrated

$$\ln G + \frac{G-1}{P_{sat}}\left(P^{+}(0) + P^{-}(L)\right) = gL. \quad (5)$$

So, one can rewrite it in terms of the output power:

$$\ln G + \frac{G-1}{GP_{sat}}\left(\frac{P^{+}_{out}}{1-R^{+}} + \frac{P^{-}_{out}}{1-R^{-}}\right) = gL. \quad (6)$$

It is known that the saturation power depends on the pump power linearly [3]. It means that the following equation holds:

$$P_{sat}\,[W] = A + B \cdot P_{pump}\,[W]\,,$$

where A and B are constants. These parameters were found by the method of least squares, and finally we obtain the following:

$$P_{sat}\,[W] = 0.0648 + 0.0642 \cdot P_{pump}\,[W]\,,$$

On figure 6 one can see the experimental and the analytically-estimated values of the saturation power. The analytical approximation and experimental results show good agreement.

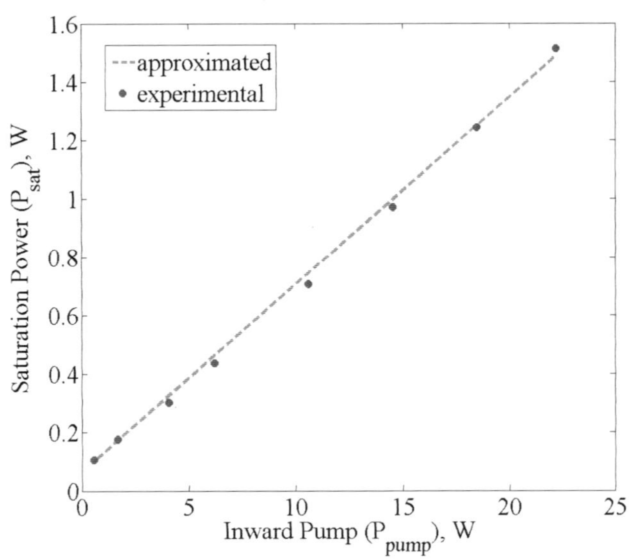

Figure 6. The saturation power computed and approximated from the experimental data.

4. Optimization

In the previous sections the total gain equation was obtained and the original system was integrated. The proposed analytical estimation has enabled to obtain all parameters of the system, and thus has made possible to perform the optimization of the output signal power. To carry out the optimization, the Split-Step Fourier method was used to solve the Schrodinger equations (equations 1) with two initial conditions: $A^{+}(t,0)$ and $A^{-}(t,0)$; from the conditions the forward and backward signals power can be derived ($P^{+}(0)$ and $P^{-}(0)$). To solve the problem one need also to know the total signal gain G, which is derived by back reflections at each side of the cavity (equation 3), and the back reflection (R^{-}) on the FBG. In this section we propose the algorithm that allows iteratively find a solution with a lower computational time spend.

The initial backward signal at the left end $\widetilde{A}^{-}(t,0)$ can be represented as a Gaussian noise with a low power. The forward signal at the left end $\widetilde{A}^{+}(t,0)$ is the reflected backward signal, therefore it can be computed using the following expression:

$$\widetilde{A}^{+}(t,0) = F^{-1}\left[f^{-}(\omega) F\left[\widetilde{A}^{-}(t,0)\right](\omega)\right],$$

where $F[A(t,z)](\omega,z)$ and $F^{-1}[A(\omega,z)](t,z)$ denote forward and backward Fourier transforms respectively, and $f^{-}(\omega)$ is the FBG spectrum reflectivity, shown on Figure 3. The backward reflection coefficient is estimated as follows:

$$R^{-} = \frac{P^{+}(0)}{P^{-}(0)} = \frac{\int \left|\widetilde{A}^{+}(t,0)\right|^{2} dt}{\int \left|\widetilde{A}^{-}(t,0)\right|^{2} dt}$$

and then the total signal gain is (equation 3)

$$G = \frac{1}{\sqrt{R^{+}R^{-}}}.$$

The next step is to find the sum $(P^{+}(0) + P^{-}(L))$ of the forward and backward signal power:

$$P^{+}(0) + P^{-}(L) = \frac{P_{sat} \cdot (gL - \ln G)}{G-1},$$

where the signal power for each direction can be computed using the following equations:

$$P^{+}(0) = \frac{P^{+}(0) + P^{-}(L)}{1 + R^{+}G}, \qquad P^{-}(L) = \frac{P^{+}(0) + P^{-}(L)}{1 + R^{-}G}.$$

After all manipulations described above the initial conditions $A^{+}(t,0)$ and $A^{-}(t,0)$ should be transformed with respect to computed signal power (the multiplication by a constant does not affect the spectrum width, that is why this transform is correct):

$$A^{-}(t,0) = \frac{P^{-}(L) \cdot G \cdot T_{R}}{\int \left|\widetilde{A}^{-}(t,0)\right|^{2} dt} \widetilde{A}^{-}(t,0), \qquad A^{+}(t,0) = \frac{P^{+}(0) \cdot T_{R}}{\int \left|\widetilde{A}^{+}(t,0)\right|^{2} dt} \widetilde{A}^{+}(t,0).$$

After the total gain is estimated, one can obtain its distribution along the z-axis, which can be found from equation 5, where the integration limit L is replaced by the variable z which is varied between 0 and L:

$$\ln\left[G(z)\right] + \frac{G(z)-1}{P_{sat}}\left(P^{+}(0) + \frac{P^{-}(0)}{G(z)}\right) = gz.$$

When the signal power stabilization process and the gain distribution computation algorithm were described, one should make the Schrodinger equations simulation process clear. As it represents the Cauchy problem, the initial conditions are set for the left side of the computational area, so the integration direction derived as left-to-right. But while the roundtrip the signal also reflects from the right side of the cavity and starts the backward propagation which direction is opposite to the direction of integration, therefore the original problem wad divided into two parts: at first the forward and the backward signals are computed in forward (left-to-right) direction (respectively to the system 1) and at the second part one considers that:

$$A^-(t, L) = R^+ \cdot A^+(t, L).$$

Then as the direction of propagation is opposite to the integration direction the signs in original system of Schrodinger equations should be changed [1], so the backward propagation system is:

$$\mp\frac{\partial A^\pm}{\partial z} = -i\beta_1\frac{\partial A^\pm}{\partial t} - i\frac{\beta_2}{2}\frac{\partial^2 A^\pm}{\partial t^2} + i\gamma\left(|A^\pm|^2 + 2|A^\mp|^2\right)A^\pm + \frac{g}{2\left(1 + P/P_{SAT}\right)}A^\pm,$$

with the initial conditions $A^+(t, L)$, $A^-(t, L)$. When one comes back to the left side of the computational area, one obtains the value:

$$\widetilde{A}^-(t, 0) = A^-(t, 2L)$$

to correct the gain distribution and the signal power. Described algorithm should be repeated iteratively until the signal spectrum and power stabilize, the stopping condition can be expressed by the following inequality:

$$\left|\left(R^-\right)_{i+1} - \left(R^-\right)_i\right| \le \epsilon.$$

Using the described method, the optimization of the output signal power as a function of the

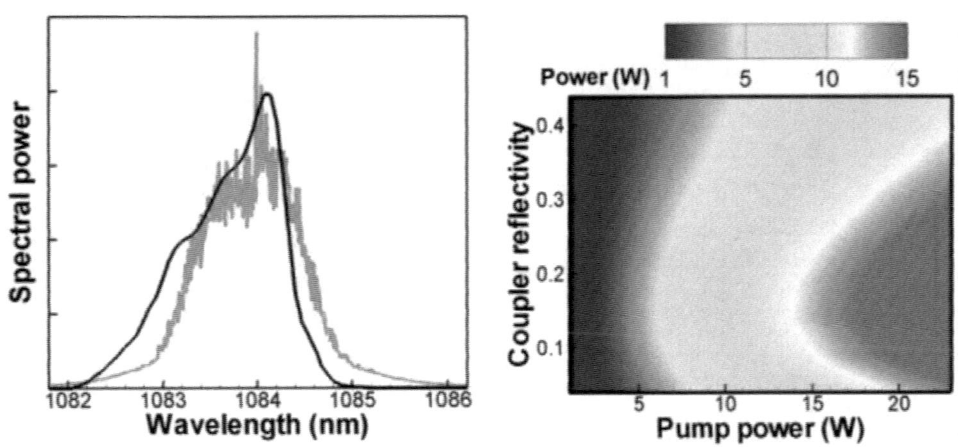

Figure 7. (left) The comparison of experimental (black curve) and numerical (blue curve) spectra, $P_{pump} = 23$W, $R^+ = 0.04$; (right) The results of optimization of the output power. The maximum power 15W is observed for low losses.

forward back reflection and the inward pump power has been carried out. These two parameters were selected, because they can be easily changed in the experiment while keeping unchanged the laser configuration. The left picture on figure 7 shows a good accordance of the computed signal spectrum to the experimental one. On the right picture the output signal power optimization results are represented. One can see that the maximum output power is observed for high pump powers and low coupler reflectivities.

5. Conclusion

In this research the one-level gain media model has been analyzed. The signal gain equation has been analytically described, and the original system has been integrated. Also the balance equation to derive the power characteristics has been obtained. The theoretical results have shown good agreement with the experimental results. This indicates that it is possible to estimate the small signal gain and the saturation power based on the experimental data. The numerical optimization based on the analytical equations and experimental data has been performed. The analysis of experimental measurements shows that the simplified model with the saturated gain is applicable to the modeling of CW strong nonlinear dynamics in the Fabry-Perot cavity.

6. Funding

The authors acknowledge the support of the Russian Scientific Foundation (project 17-71-20082).

References

[1] **S. K. Turitsyn, A. E. Bednyakova, M. P. Fedoruk, A. I. Latkin, A. A. Fotiadi, A. S. Kurkov, and E. Sholokhov.** Modeling of CW Yb-doped fiber lasers with highly nonlinear cavity dynamics. Optics Express Vol. 19, No. 9, 25 April 2011.

[2] **C. Barnard, P. Myslinski, J. Chrostowski, and M. Kavehrad.** Analytical model for rare-earth-doped fiber amplifiers and lasers. IEEE J. Quantum Electron. 30, 18171830 (1994).

[3] **O. V. Shtyrina, A. V. Ivanenko, I. A. Yarutkina, A. V. Kemmer, A. S. Skidin, S. M. Kobtsev and M. P. Fedoruk.** Experimental measurement and analytical estimation of the signal gain in an Er-doped fiber. Journal of the Optical Society of America B Vol. 34, No. 2, February 2017.

[4] **A. E. Siegman**, Lasers (University Science Books, 1986).

[5] **A. S. Kurkov and E. M. Dianov.** Moderate-power CW fiber lasers. Quantum Electron. 34(10), 881900 (2004).

[6] **B. G. Bale, O. G. Okhotnikov, and S. K. Turitsyn.** Modeling and technologies of ultrafast fiber lasers. In Fiber Lasers, O. G. Okhotnikov, ed. (Wiley, 2012).

[7] **S. K. Turitsyn.** Theory of energy evolution in laser resonators with saturated gain and non-saturated loss. Opt. Express 17, 1189811904 (2009).

[8] **I. A. Yarutkina, O. V. Shtyrina, A. Skidin, and M. P. Fedoruk.** Theoretical study of energy evolution in ring cavity fiber lasers. Opt. Commun. 342, 2629 (2015).

[9] **T. Pfeiffer and H. Bullow.** Analytical gain equation for erbium-doped fiber amplifiers including mode field profiles and dopant distribution. IEEE Photon. Technol. Lett. 4, 449451 (1992).

Pressure shift coefficient measurements in an RF discharge for Ar 4s[3/2]$_2$—5p[3/2]$_3$ transition with the help of diode-laser absorption spectroscopy

A K Chernyshov [1,2], P A Mikheyev [1,2], N N Lunev [1], and V N Azyazov [1,2]

[1] Lebedev Physical Institute, 443011, Samara, Russia
[2] Samara National Research University, 443086, Samara, Russia

e-mail: chak@fian.smr.ru

Abstract. Optically pumped all-rare-gas laser (OPRGL) with unique properties were recently proposed with a possibility to obtain the laser power on the order of hundreds of Watts from a cubic centimeter. To provide high laser efficiency, the pumping radiation has to match the absorption spectrum of the rare gas metastables. To meet this condition a reliable diagnostics of the key parameters of the active medium is required and knowledge of the broadening and shift coefficients for corresponding transitions of rare gases is necessary. In this paper, the diode-laser absorption spectroscopy was employed to determine the pressure shift coefficient for 811.5 nm Ar line. The value of obtained coefficient in pure argon reduced to 300 K is $-(2.1 \pm 0.1) \times 10^{-10}$ s^{-1}cm^3. In the course of the study the pressure broadening coefficient was also evaluated and found to be $(2.4 \pm 0.5) \times 10^{-10}$ s^{-1}cm^3.

Keywords: Optically pumped all-rare-gas laser, metastable argon atoms, radio-frequency discharge, diode-laser absorption spectroscopy, spectral line pressure shift

1. Introduction

The well-known diode-pumped alkaline-vapour lasers (DPAL) attracted considerable attention because of their possibility to obtain optical power at a level of several kilowatts from a volume of a few tens of cubic centimetres [1]. At the same time, these lasers possess intrinsic technical difficulties associated with high reactivity of the alkali metals. As a chemically passive alternative to DPAL, in 2012 lasing was demonstrated employing metastables of the heavier rare gases (neon argon, krypton and xenon) [2, 3]. In such systems, both electric discharge and radiation from high-power diode lasers are supposed to be used. In this case, the electric discharge provides the population of the lower (metastable) laser level, the optical pumping populates the upper pump level and subsequent collisional energy transfer populates the upper laser level. These lasers are now called optically pumped rare gas lasers (OPRGL), they operate according to the traditional three-level scheme.

A high lasing threshold characterizes the three-level laser scheme. Model estimates of an OPRGL show that to obtain a specific optical power of the order of 100 W/cm^3, it is necessary that the number density of metastable atoms to be on an order of 10^{12}-10^{13} cm^{-3} at a gas mixture pressure of about 1 atm [4]. In addition, the radiation of diode lasers providing optical pumping must be spectrally

IOP Conf. Series: Journal of Physics: Conf. Series **999** (2018) 012010 doi:10.1088/1742-6596/999/1/012010

matched with the absorption line of the rare gas metastables. To meet these conditions the reliable diagnostics of the laser active medium is needed. It means, first of all, that the key parameters such as metastables' number density and the pressure broadening/shift coefficients for the transitions of interest are known.

Earlier, we measured the pressure-broadening coefficients of absorption lines of argon and krypton atoms with the help of diode-laser absorption spectroscopy [5-7]. In our experiments, the metastable atoms were generated in plasma of an RF discharge for gas mixtures that are commonly used for OPRGL. In this paper, the previously used spectroscopic setup was upgraded, which allowed us to measure the pressure broadening and the shift coefficients of absorption lines in rare gases simultaneously. The above mentioned coefficients for the $4s[3/2]_2 \rightarrow 5p[3/2]_3$ transition of the Ar atom were measured upon excitation by a 40 MHz RF discharge in pure argon.

2. Description of the method

Dependence of the shift of an absorption line $\Delta\nu_{sh}$ on pressure P (Torr) and temperature T(K) is as follows [8]

$$\Delta\nu_{sh}(P,T) = 3.22\cdot10^{16}\gamma(T_0)\left(\frac{T_0}{T}\right)^{0.7}P. \tag{1}$$

Here T_0 is the reference temperature chosen equal to 300 K. As follows from (1), to determine the shift coefficient $\gamma(T_0)$ it is necessary to measure three quantities, namely the shift of the spectral line $\Delta\nu_{sh}$, gas temperature T and pressure P in a gas discharge cell. It is easy to measure gas pressure P by means of the commercially available device and it is the most reliably determined quantity from the specified above parameters set.

In usual spectroscopic examinations gas temperature T in a sample cell is equal to temperature of the ambient air and changes in time only slightly, therefore this parameter does not create any problems during measurements. On the other hand, in plasma spectroscopy, gas temperature in a discharge noticeably differs from ambient air temperature and can vary considerably owing to changes of excitation conditions in an RF discharge. For example, gas temperature T in a discharge can vary considerably with changes in pressure P and that it is necessary to consider.

The temperature of a neutral gas T in plasma is convenient to determine by non-invasive probing from a profile of the spectral line. For this purpose, the measured profile of the absorption line was fitted numerically by the Voigt function, which is a convolution of the Doppler and Lorentz contours. Numerical fitting to the Voigt function allowed to reconstruct the widths (FWHM) of the Doppler *WD(T)* and the Lorentz *WL(P, T)* contours simultaneously in accordance with the technique [6, 7]. In turn, the width *WD(T)* of the Doppler contour is uniquely determined by the gas temperature T in accordance to the known expression [9]

$$WD(T) = 7.16\cdot10^{-7}\nu_0\left(\frac{T}{M_r}\right)^{1/2}, \tag{2}$$

where M_r is the mass of the Ar atom in atomic units. Then for the reference temperature T_0, the width of the Doppler contour of argon will be equal to $WD(T_0) = 0.724$ GHz. Using equation (2), expression (1) for the line shift $\Delta\nu_{sh}$ can be rewritten in another form

$$\Delta\nu_{sh}(P,T)\left(\frac{WD(T)}{0.724}\right)^{1.4} = 3.22\cdot10^{16}\gamma(T_0)P. \tag{3}$$

As a result of such a transformation, a combination of experimentally measured quantities on the left-hand side of relation (3), exhibits linear dependence on gas pressure P. The slope of the line given by the expression (3) is related to the pressure shift coefficient $\gamma(T_0)$.

IOP Conf. Series: Journal of Physics: Conf. Series **999** (2018) 012010 doi:10.1088/1742-6596/999/1/012010

The dependence of the width of the Lorentz contour *WL(P, T)* on the temperature *T* and pressure *P* is similar to that for the line shift $\Delta\nu_{sh}(P,T)$ [6]

$$WL(P,T)\left(\frac{WD(T)}{0.724}\right)^{1.4} = 3.22\cdot 10^{16}\cdot 2\xi(T_0)P\,. \tag{4}$$

Since the coefficient $\xi(T_0)_{\text{Ar-Ar}}$ for argon was previously measured [5, 6], it is possible to determine the gas pressure *P* from the value *WL(P,T)* and compare it with the measured value and thus, to verify the correctness of all intermediate actions associated with the definition of *WD(T)* and *WL(P,T)*.

3. Experimental setup

Diode-laser spectroscopy was used to measure the shape of the argon line. A simplified diagram of the experimental setup, which consisted of three main parts, is shown in figure 1. First, it had a sample cell to which RF electronic units and a rare gas pumping system were connected. The Ar-sample cell is a quartz discharge tube 50 cm long, 15 mm in diameter with a wall thickness of 1.5 mm. The glass windows were fixed at an angle to optical axis at the ends of the discharge tube. External electrodes were attached at the middle of the sample cell, as shown in the inset of figure 1. The electrodes were U-shaped pieces of thick copper wire with 10 cm sides. A symmetrical 2-wire line was connected to the electrodes, through which the RF power (~10 W) from the commercially available 40 MHz generator reached the discharge cell through a matching circuit (not shown).

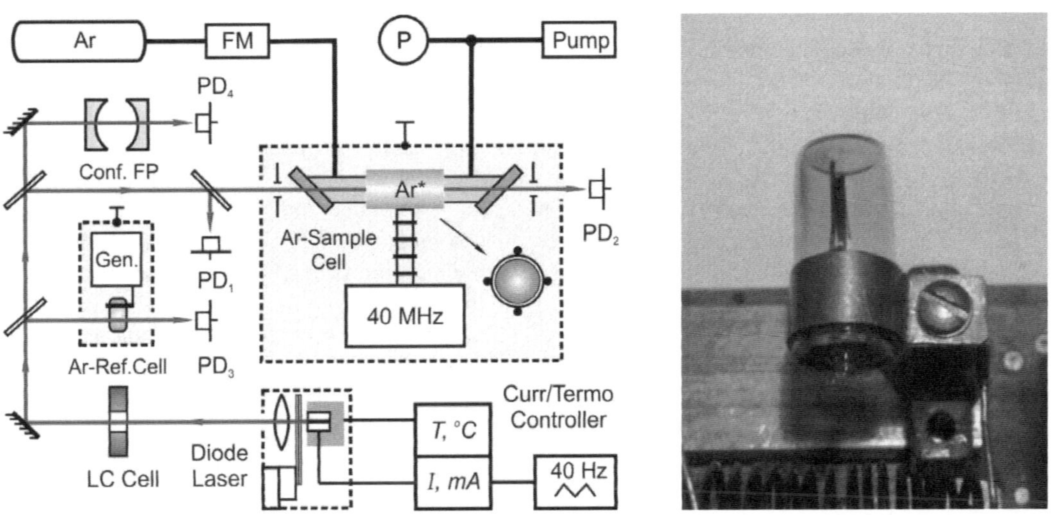

Figure 1. Simplified diagram of the experimental setup (on the left) and the fluorescent lamp starter used as a reference cell (on the right). The cross section of the discharge tube with four wire electrodes is shown near to the sample cell within the grounded chamber.

The experiments were performed in the regime of constant gas pumping. Argon (0.999 purity) from the gas cylinder through the reduction gear and the mass flow controller FM (Bronkhorst) was fed to the inlet socket of the cell. The outlet socket of the sample cell was connected to an oil-free vacuum pump (nXDS10i, Edwards) through a needle choke valve. The gas pressure in the discharge tube was regulated by the needle valve and controlled by a capacitive pressure sensor P (Metran) installed near the outlet.

The second component is the source of probing optical radiation, which is a diode laser (DL) L808PO30 (Thorlabs), equipped with a short external resonator of the original configuration [10, 11]. The current/temperature controller ITC4001 (Thorlabs) was used to power the laser and stabilize its

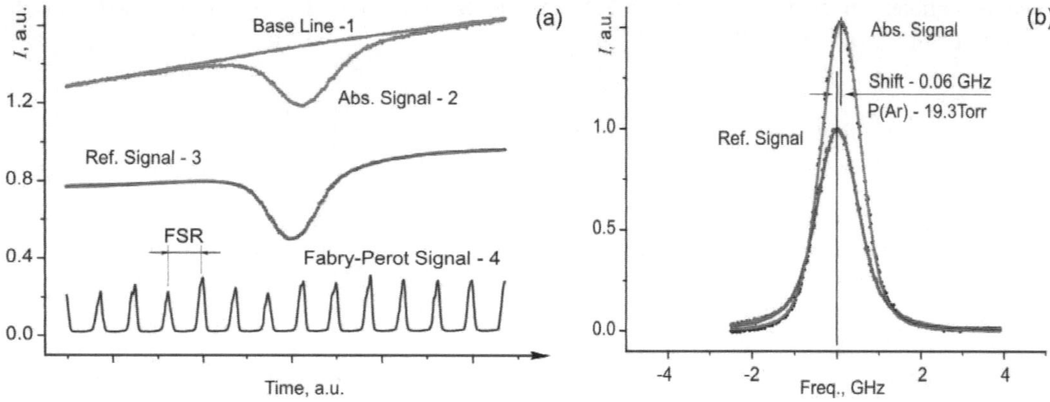

Figure 2. Observation of 811.5 nm line in argon: a) the initial absorption signals with the tilted baseline after passing the appropriate cells and the transmission resonance signals of the reference interferometer (FSR = 750 MHz) recorded versus time; b) Processed absorption signals with a horizontal baseline, as a function of detuning from the atomic line center.

temperature. A triangular signal from an external generator with a frequency of 40 Hz was added into the injection current of the DL through the controller ITC4001 to obtain continuous tuning of the laser frequency. Liquid crystal cell (LC), operating in the quarter wave plate mode, acted as a polarization isolator [12]. The spectrum monitoring system of the laser radiation included a spectrometer (not shown in figure 1) AvaSpec-3648 (Avantes) and a confocal interferometer (F $\approx$ 200, FSR = 750 GHz). With their help, it was found that the range of the discrete temperature tuning of the laser wavelength was ~3nm, the continuous current tuning (near 810 nm) of the laser frequency was on the order of 36 GHz. The spectral width of the probing radiation was estimated to be 50 MHz. The maximum optical output power available from the laser diode was up to 5 mW, which was measured with the help of a photodiode power meter (PD300, Ophir).

The third component of the experimental setup was a small sealed-off discharge cell, which was used as a reference for the optical frequency, relative to which the shift of the argon absorption line in the sample cell was determined. It is rather difficult to manufacture a high-quality sealed gas-discharge cell in an ordinary optical laboratory [13]. This is due to the fact that the accumulated concentration of metastables of the rare gas is very sensitive to the presence of contaminants. Therefore, the Ar starter from the fluorescent lamp (S2 220/240V, 4-22W, Philips) was used as the reference cell [14]. To excite the radio-frequency discharge in the glass bulb of the starter, instead of the built-in electrodes, an external ring electrode was used, which simultaneously served as the holder figure 1 (on the right). This ring electrode was connected directly to a low-power 50 MHz transistor generator. The role of the second electrode was performed by the grounded metal case of the device. A thin probing laser beam passed through the starter bulb between the internal electrodes and was registered with a photodetector. Silicon monolithic photodiodes with on-chip transimpedance amplifiers (OPT101, Texas Instruments) were used as photodetectors in the setup.

4. Results and discussion

Figure 2(a) shows the initial signals recorded in the experiment. The sloped line 1 is the base line of zero absorption, recorded by the photodetector PD1 which was installed at the entrance to the Ar-sample cell. The slope of line 1 is caused by a change in the output power of the diode laser when its frequency is tuned by ramping of injection current. Line 2 is the absorption signal recorded by PD2 at the output of the discharge cell with argon. Line 3 is the signal from PD3 and corresponds to argon absorption in the reference cell.

IOP Conf. Series: Journal of Physics: Conf. Series **999** (2018) 012010 doi:10.1088/1742-6596/999/1/012010

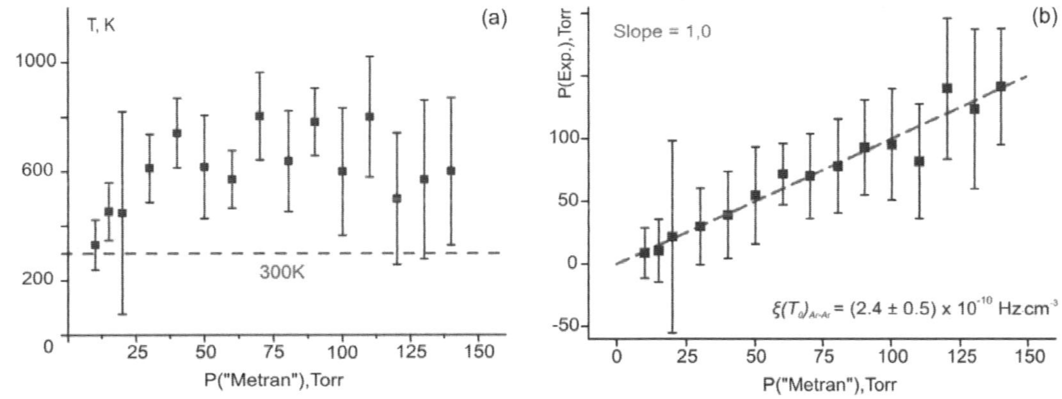

Figure 3. The results of measurements of temperature (a) and pressure (b) in the sample cell at different argon pressures.

Finally, line 4 in figure 2(a) shows the transmission resonances of a Fabry-Perot interferometer with a free spectral range of 750 MHz between the peaks. The distance between the peaks of the resonances determine the frequency range, which was used for a linearization procedure of the scale along the horizontal axis. After elimination of the baseline slope by dividing signal 1 to signal 2, followed by taking the logarithm the normalized spectral profile was obtained, as shown by points in figure 2 (b). These experimental points were fitted to the Voigt profile using the Origin software. In the range of gas pressures from 20 to 100 Torr and moderate temperatures, the *WD(T)* and *WL(P,T)* widths are comparable. In this case, fitting allows to determine values of *WD(T)* and *WL(P,T)* from the Voigt profile accurately.

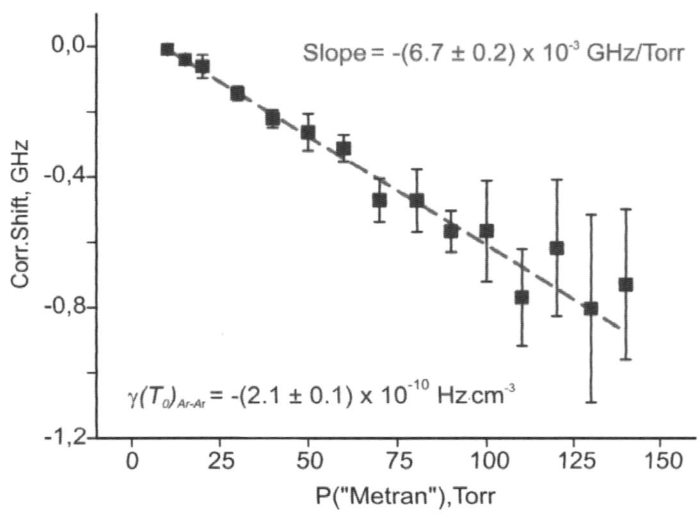

Figure 4. The results of measurement of the corrected center shift of the 811.5 nm line from the argon pressure in the sample cell

Gas temperature *T* averaged along the laser beam path was determined using expression (2) and the corresponding widths of the Doppler contours *WD(T)*. As it was shown in Ref. [6] the left part of relation (4) exhibits weak dependence on temperature gradient along the beam. The graph in figure 3(a) shows the temperature *T* as a function of the gas pressure *P* in the sample cell. It can be seen that the temperature of the gas varies from point to point. This is because during the pressure change in the sample cell it was necessary to adjust the matching circuit between the RF generator and the discharge. The calculated values also confirm that the temperature of neutral gas in the RF discharge is larger than room temperature (300 K). This is a convenient criterion for data verification.

The *WL(P,T)* and *WD(T)* widths make it possible, in accordance with formulas (2) and (4), to estimate the gas pressure P_{exp} in the sample cell. If the values of P_{exp} are put on a graph against the measured values of P obtained from the pressure sensor *Metran*, then in the ideal case, the points must lie on the line with the unit slope coefficient and passing through the origin point of coordinate system (dotted line in figure 3(b)). This condition makes it possible to determine the value of the pressure broadening coefficient of the spectral line $\xi(T_0)$. Since, if an incorrect $\xi(T_0)$ value is used, then the points of the graph will not be approximated by a straight line that is in good agreement with the dotted line with a unit slope coefficient. In this work, the experimental points in figure 3(b) were fitted straight line with a slope coefficient equal to 1.00 ± 0.06, while the value of $\xi(T_0)$ was chosen equal to $(2.4 \pm 0.5) \times 10^{-10}$ $s^{-1}cm^3$. For comparison, the value of the same coefficient obtained by us in a somewhat different way [5, 6] was $\xi(T_0)_{Ar\text{-}Ar} = (2.85 \pm 0.1) \times 10^{-10}$ $s^{-1}cm^3$. Within the error limits, both values coincide.

Using the Doppler width *WD(T)*, it is possible to correct the temperature contribution to the shift $\Delta\nu$ of the spectral line. In accordance with (3), if the corrected shift $\Delta\nu$ is plotted along the vertical axis, and the gas pressure P in the sample cell along the horizontal axis, a straight line should be obtained again. The experimental points are well approximated by a straight line as shown in figure 4. The error of the spectral line shift increases with pressure in the discharge. This is because the profile width of the investigated Ar line also increases with pressure in the discharge cell, which makes it difficult to determine the central frequency. However, the values of the slope coefficient defined either for the entire dataset or only for the points with small error are identical and equal to $-(6.7 \pm 0.2) \times 10^{-3}$ GHz/Torr. The value of the coefficient of pressure shift determined by the slope of the graph in figure 4 was $\gamma(T_0)_{Ar\text{-}Ar} = -(2.1 \pm 0.1) \times 10^{-10}$ $s^{-1}cm^3$. It is well known that the ratio $\gamma(T_0)/(2\xi(T_0))$ must be equal to 0.368 [15]. The ratio of these coefficients defined in the work, gives a value of 0.44 ± 0.09, which is in good agreement with the theoretical estimate.

5. Conclusion

In this paper the method is proposed for simultaneous measurement of the pressure broadening and shift coefficients of the spectral lines, which makes it easy to take into account the contribution from the gas temperature. The broadening and shift coefficients for the $4s[3/2]_2 \rightarrow 5p[3/2]_3$ Ar transition at 811.5 nm, which are important for the development of promising OPRGL lasers, were determined. Additionally it was demonstrated that easily accessible starters of fluorescent lamps can be used as sealed discharge cells for reference points of optical frequency.

Acknowledgments

This work was supported by the Ministry of Education and Science of the Russian Federation within the framework of the State Order under Project 3.1715.2017/4.6.

References

[1] Krupke W F 2012 *Progress in Quantum Electronics* **36** 4
[2] Han J, Heaven M C 2012 *Opt. Lett.* **37** 2157
[3] Han J, Glebov L, Venus G and Heaven M C 2013 *Opt. Lett.* **38** 5458
[4] Demyanov A V, Kochetov I V and Mikheyev P A 2013 *J. Phys. D: Appl. Phys.* **46** 375202
[5] Mikheyev P A, Chernyshov A K, Ufimtsev N I and Vorontsova E A 2015 *Proc. SPIE* **9255** 92552W
[6] Mikheyev P A, Chernyshov A K, Ufimtsev N I, Vorontsova E A and Azyazov V N 2015 *J. Quant. Spectrosc. Radiat. Transt.* **146** 1

IOP Conf. Series: Journal of Physics: Conf. Series **999** (2018) 012010 doi:10.1088/1742-6596/999/1/012010

[7] Gildina A R, Mikheyev P A, Chernyshov A K, Ufimtsev N I and Azyazov V N 2017 *Photonics* **65 (5)** 44
[8] Belostotskiy S G, Donnelly V M, Economou D J and Sadeghi N 2009 *IEEE Trans. Plasma Sci.* **37** 852
[9] Demtroder W 2003 *Laser Spectroscopy: basic concepts and instrumentation* (Berlin, Heidelberg, New York: Springer) chapter 3 pp 68–72
[10] Chernyshov A K, Chernyshova E A 2011 *Physics of Wave Phenomena* **19** 89
[11] Chernyshov A K, Mikheyev P A 2016 *Proc. SPIE* **10176** 101760T
[12] Chernyshov A K, Kotova S P 2006 *Instrum. Exp. Tech.* **49** 92
[13] Sukenik C I, Busch H C 2002 *Rev. Sci. Instrum.* **73** 493
[14] Chernyshov A K 2018 *Instrum. Exp. Tech.* **61** 141
[15] Moussounda P S, Ranson P 1987 *J. Phys. B: At. Mol. Phys.* **20** 949

IOP Conf. Series: Journal of Physics: Conf. Series **999** (2018) 012011 doi:10.1088/1742-6596/999/1/012011

Spectrum Evolution of Accelerating or Slowing down Soliton at its Propagation in a Medium with Gold Nanorods

Vyacheslav A. Trofimov, Tatiana M. Lysak

Lomonosov Moscow State University, Moscow 119992, Russia

E-mail vatro@cs.msu.ru

Abstract. We investigate both numerically and analytically the spectrum evolution of a novel type soliton - nonlinear chirped accelerating or decelerating soliton - at a femtosecond pulse propagation in a medium containing noble nanoparticles. In our consideration, we take into account one- or two-photon absorption of laser radiation by nanorods, and time-dependent nanorod aspect ratio changing due to their melting or reshaping because of laser energy absorption. The chirped solitons are formed due to the trapping of laser radiation by the nanorods reshaping fronts, if a positive or negative phase-amplitude grating is induced by laser radiation. Accelerating or slowing down chirped soliton formation is accompanied by the soliton spectrum blue or red shift. To prove our numerical results, we derived the approximate analytical law for the spectrum maximum intensity evolution along the propagation coordinate, based on earlier developed approximate analytical solutions for accelerating and decelerating solitons.

1. Introduction

Noble metal nanoparticles reveal the unique optical properties, allowing their use in devices of ultrahigh density information storage [1-6]. The optical properties of metal nanorpatrticles are based on their high sensitivity to the wavelength and polarization of laser radiation. That is why the influence of the noble metal nanoparticle aspect ratio and orientation, and local environment on the nanoparticle optical response, is widely investigating nowadays [7-14].

Nanoparticles aspect ratio changing because of photo-thermal reshaping [15-19] provides the ability for five-dimensional recording [1-6]. Further development of five-dimension memory can be related to the information density increase, for example, due to the decrease of the space between the recording layers. In our opinion, the solution of this problem may be related to the self-similar mode of laser radiation propagation in the medium containing noble metal nanorods. Indeed, nanoparticles reshaping due to the laser radiation absorption leads to the laser pulse spectrum changing caused by the pulse chirping. These spectrum distortions can result in violation of optimal conditions for information recording, as well as (that is more essential) in the false information recording and reading.

In our previous papers [20-25] we investigated a femtosecond pulse propagation in a medium containing nanorods under conditions of non-instantaneous absorption due to the nanorods reshaping and dependence of laser radiation absorption on nanoparticle aspect ratio. In particular, we investigated accelerating and slowing down soliton mode of laser radiation propagation in the medium containing nanorods under one- and multi-photon absorption [21- 24]. In papers [22] and

[23], we derived the approximate analytical solutions for these modes of laser radiation propagation, and compared these solutions with the computer simulation results. In numerical simulations, red or blue shift of the pulse spectrum was observed for accelerating or slowing down soliton propagation, respectively.

In present paper, we use the approximate analytical solutions, derived in [23] and [24] for accelerating and decelerating soliton modes of laser radiation propagation in the medium containing nanorods, to develope the analytical law for the soliton spectrum evolution observed in computer simulations.

2. Problem statement

We consider a femtosecond laser pulse propagation in a medium with nanorods, taking into account the nanorods aspect ratio changing due to their reshaping because of laser energy multi-photon absorption (MPA). In the framework of slowly varying envelope of wave packet, this process can be described by the following dimensionless nonlinear equations [21]:

$$\frac{\partial A}{\partial z}+iD\frac{\partial^2 A}{\partial t^2}+(\varepsilon-1)\left(\delta_0+i\xi\right)|A|^{2(k-1)}A=0, \tag{1}$$

$$\frac{\partial \varepsilon}{\partial t}=-\tilde{\delta}(\varepsilon-1)|A|^{2k} \tag{2}$$

with initial and boundary conditions for complex amplitude and aspect ratio of nanorods

$$A(z,t)\big|_{t\to\pm\infty}=0,\quad \left.\frac{\partial A(z,t)}{\partial t}\right|_{t\to\pm\infty}=0,\quad A(0,t)=A_0(t),\quad \varepsilon(z,t=0)=\varepsilon_0,\quad 0\le z\le L_z.$$

Above A is a dimensionless slowly varying envelope of a wave packet normalized on a square root from the maximum incident pulse intensity ($z=0$). $\varepsilon=a/b$ is a nanorod aspect ratio, where a and b are major and minor axes of nanorods, ε_0 is its initial value. k is the number of photons involved the the absorption: $k=1$ corresponds to one-photon absorption, $k=2$ - to two-photon absorption. Variable z is a dimensionless longitudinal coordinate, along which the optical radiation propagates; t is a dimensionless time in the system of coordinates moving with a pulse, time is measured in the units of τ_{pulse} - duration of the incident pulse; L_t is a dimensionless time interval, during which the laser pulse interaction with nanorods is analyzed, L_z is a dimensionless length of nonlinear medium. Parameter D characterizes the group velocity dispersion (GVD). Parameters δ_0 and $\tilde{\delta}$ characterize the absorption of laser light and a part of absorbed energy spent on nanorods reshaping, correspondingly. Coefficient ξ characterizes the laser pulse self-action due to the wave packet carrier frequency detuning from the central frequency of the nanorod absorption spectrum. The case of $\xi=0$ corresponds to an optical pulse propagation in a medium with pure amplitude grating. It means an influence only of a laser energy absorption on the laser pulse propagation. In the opposite case ($\xi\neq 0$), the phase grating is also induced by the laser radiation. It should be mentioned, that the positive sign of the parameter ξ (this case is named by us as positive grating) corresponds to pulse compression and the laser pulse decompression occurs at negative sign of this parameter (this case is named by us as negative grating).

The function $f(\varepsilon)$ describes the dependence of TPA process on the nanorod aspect ratio ε . The Boyd's and Shen's shape factor expression [26] for the absorption coefficient of gold nanorods, calculated in the dipole approximation [27] for physical parameters $\varepsilon_1=-22.4$, $\varepsilon_2=1.8$ for Au at $\lambda=800$ and ε_m=3, gives the following dependence

$$f(\varepsilon)=\frac{\left(\varepsilon_2/A^2\right)}{\left(\varepsilon_1+\dfrac{1-A}{A}\varepsilon_m\right)^2+\varepsilon_2^2}, \quad (3)$$

$$A=\left[1-\frac{\xi Q_1'(\xi)}{Q_1(\xi)}\right]^{-1}, \quad \xi=\left[1-\left(\frac{1}{\varepsilon}\right)^2\right]^{-1/2}, \quad Q_1(\xi)=\left(\frac{\xi}{2}\right)\ln\left[\frac{\xi+1}{\xi-1}\right]-1, \quad Q_1'(\xi)=\frac{dQ_1(\xi)}{d\xi}, \quad (4)$$

In equation (3), ε_m is the dielectric permittivity of the ambient medium, $\varepsilon_1+i\varepsilon_2$ is the complex dielectric permittivity of the nanorods. Below we consider the linear approximation of this dependence:

$$f(\varepsilon)=\varepsilon-1 \quad (5)$$

The linear dependence (3) adequately approximates dependence (3)-(4) (figure1, red line) if the aspect ratio varies from 1 to 2. Below we choose initial value of aspect ratio as $\varepsilon_0=2$.

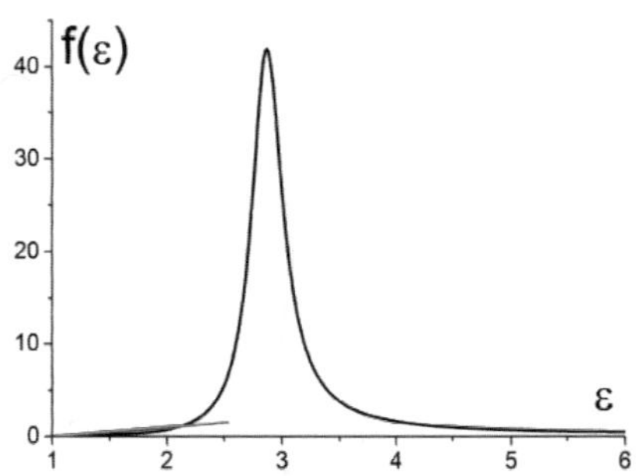

Figure 1. Dependence $f(\varepsilon)$ defined by formulas (3)-(4) for gold nanorods at the falling radiation with $\lambda=800$ nm (black line), or by formula (5) (red line).

If the aspect ratio varies from 1 to 2, then a laser radiation interaction with nanorods occurs far from the nonlinear absorption resonance. In this case, a linear dependence (5) (figure 1, red line) adequately approximates dependence (3)-(4) (figure 1, black line). Below we choose the following initial value of aspect ratios $\varepsilon_0=2$.

We specify the incident Gaussian pulse

$$A(z=0,t)=A_0(t)=\exp\left(-((t-L_t/2)/\tau)^2\right),$$

τ is a dimensionless pulse duration.

We follow the laser pulse centre position along the z-coordinate

$$\tau_c(z)=\int_0^{L_t}(t-L_t/2)|A(z,t)|^2\,dt \Big/ \int_0^{L_t}|A(z,t)|^2\,dt$$

and the maximum position for the laser pulse spectra along the z-coordinate.

$$\omega_{\max}(z)=\arg\max_{\omega}|A(z,\omega)|^2,$$

which means that, for a fixed value of the variable z, the function $|A(z,\omega)|^2$ takes the maximum value for $\omega=\omega_{\max}(z)$.

3. Analytical consideration

In our previous papers [21-23], we investigated numerically and analytically the accelerating and slowing down solitons for the problem under consideration. In particular, we derived the approximate analytical solution, which describes the soliton propagation, and computed the soliton center shift along the z-coordinate $\tau_c(z)$. In our investigations, the obvious shift of the soliton spectrum into the area of lower or higher frequencies was clearly seen. Present research is devoted to the derivation of the analytical expression for the maximum spectrum shift along the z-coordanate $\omega_{\max}(z)$.

In [22,23], we considered the following approximation of the accelerating or slowing down soliton

$$A(z,t) = B(z)ch^{-1}(\zeta)\exp(-is(z,\zeta)),\ \varepsilon(z,t) = \varepsilon(z,\zeta),\ \zeta = \left(t - t_c(z)\right)/\tau_s(z), \quad (6)$$

$$s(z,\zeta) = a(z) + b(z)\zeta + h(z)\ln(ch(\zeta)) \quad (7)$$

$$B(0) = B_0,\ t_c(0) = t_0,\ \tau_s(0) = \tau_{s0}. \quad (8)$$

Here $A(z,t)$ is the soliton complex amplitude, functions $B(z)$, $s(z,\zeta)$, $t_c(z)$ and $\tau_s(z)$ describe the soliton amplitude, phase distribution, and pulse center and its duration, respectively, with their initial values B_0, t_0 and τ_{s0} for the time moment $t=0$. Parameters a_0, b_0 and h_0 define the incident pulse phase distribution. Function $a(z)$ describes a soliton phase shift along the z-coordinate, and is unchangeable in time. Functions $b(z)$, $h(z)$ describe the soliton frequency evolution along the propagation coordinate and a pulse chirp evolution, respectively.

Note that, as it follows from our approximate analytical solution for accelerating and decelerating soliton, the function $h(z)$ does not depend on the z-coordinate, so that $h(z) \equiv h_0$ [22, 23]. In this case, one can write the following sequence of equalities for the Fourier transform $A(z,\omega)$ of the soliton complex amplitude $A(z,t)$:

$$\begin{aligned}
A(z,\omega) &= \frac{1}{\sqrt{2\pi}} \int_{-\infty}^{\infty} B(z)ch^{-1}(\zeta)\exp(-is(z,\zeta))\exp(i\omega t)dt = \\
&= \frac{B(z)}{\sqrt{2\pi}}\exp(-ia(z)) \int_{-\infty}^{\infty} ch^{-1}(\zeta)\exp(-i(b(z)\zeta + h_0\ln(ch(\zeta)))\exp(i\omega t)dt = \\
&= \frac{B(z)}{\sqrt{2\pi}}\tau_s(z)\exp\left(-i\left(a(z) - \omega t_c(z)\right)\right)\times \\
&\times \int_{-\infty}^{\infty} ch^{-1}(\zeta)\exp(-i(b(z)\zeta + h_0\ln(ch(\zeta)))\exp(-i\omega\tau_s\zeta)d\zeta = \\
&= \frac{B(z)}{\sqrt{2\pi}}\tau_s(z)\exp\left(-i\left(a(z) - \omega t_c(z)\right)\right) \int_{-\infty}^{\infty} ch^{-1-ih_0}(\zeta)\exp(i(\omega\tau_s(z) - b(z))\zeta)d\zeta
\end{aligned} \quad (9)$$

Using (9), the spectrum intensity can be computed as

$$\left|A_\omega(z,\omega)\right|^2 = B^2(z)\tau_s^{\ 2}(z)F(\omega\tau_s(z) + b(z)),\quad F(x) = \left|\frac{1}{\sqrt{2\pi}}\int_{-\infty}^{\infty} ch^{-1-ih_0}(\varsigma)\exp(-ix\varsigma)d\varsigma\right|^2 \quad (10)$$

The maximum spectrum intensity position for the given medium section $z = const$ is determined from (10) by the position of the function $F(x)$ maximum

$$\omega_{\max}(z)\tau_s(z)-b(z)=\arg\max_x F(x). \tag{11}$$

Eq.(11) means that the function $F(x)$ takes the maximum value for $x=\omega_{\max}(z)\tau_s(z)-b(z)$.

Taking into account that the position of the maximum of the function $F(x)$, determined by (10), does not depend on the z-coordinate, the following equality is valid

$$\omega_{\max}(z)\tau_s(z)-b(z)=\arg\max_x F(x)=\omega_{\max}(0)\tau_s(0)-b(0). \tag{12}$$

It is possible to obtain the dependence of the maximum spectra intensity position $\omega_{\max}(z)$ on the z-coordinate from equality (12) if the position $\omega_{\max}(0)$ of the maximum of the incident pulse spectrum intensity is known, as well as the dependences $\tau_s(z)$ and $b(z)$.

Obviously, the maximum spectra position of the incident Gaussian pulse spectrum in the problem under consideration is

$$\omega_{\max}(0)=0$$

In our approximation, the pulse duration is a constant value: $\tau_s(z)=\tau_{s0}$ [22, 23]. So, equation (12) determines the sought-of dependence for the maximum intensity position of the soliton spectrum:

$$\omega_{\max}(z)=\frac{b(z)-b(0)}{\tau_{s0}}. \tag{13}$$

Equation (13) shows that the soliton maximum spectrum intensity position does not depend on the soliton amplitude $B(z)$ or the soliton center $t_c(z)$, but it is determined by the pulse duration τ_{s0} and the evolution $b(z)$ of the soliton frequency along the propagation coordinate.

4. Computer simulation results

This section is devoted to the comparison of the soliton maximum spectrum intensity position evolution given by analytical prediction (13) and obtained in computer simulation. In section 4.1, this comparison is provided for decelerating soliton in the medium containing nanorods under one-photon absorption. In section 4.1, we compare the analytical and computer simulation results for accelerating soliton in the medium containing nanorods under two-photon absorption.

4.1. Deceleration soliton

In paper [22], we have shown that the decelerating soliton is formed in the medium containing nanorods under one-photon absorption and negative phase grating for parameters $\delta_0=0.2$, $\xi=-5$, $\tilde{\delta}=5$, $D=0.1$ (figure 2). The soliton shape of the deceleration soliton is confirmed by its spectrum evolution: during the decelerating soliton propagation, the soliton spectrum preserves its shape [22] and shifts into the area of higher (slower) frequencies.

In our approximation (6)-(8), the soliton centre evolution (figure 2b) is described by the quadratic dependence

$$\begin{aligned} t_c(z)=t_0+\left(2\frac{D}{\tau_{s0}}b_0+\delta_0(\varepsilon_0-1)\tilde{\delta}\tau_{s0}{}^2B_0{}^2exp\left(-\tilde{\delta}B_0{}^2\tau_{s0}\right)\right)z-\\ -D\left(\xi-h\cdot\delta_0\right)(\varepsilon_0-1)\tilde{\delta}B_0{}^2exp\left(-\tilde{\delta}B_0{}^2\tau_{s0}\right)z^2 \end{aligned} \tag{14}$$

Dependence (14) was obtained in paper [22] under assumption that the soliton amplitude remains the same for a large trace of propagation (it takes place, for example, if laser energy depletion is weak). This assumption is valid for the considered case. Under this assumption, the evolution of the soliton frequency along the z-coordinate is a linear function [22]

$$b(z) = b_0 - \left(\xi - h\cdot\delta_0\right)(\varepsilon_0 - 1)\tilde{\delta}\tau_{s0}B^2 exp\left(-\tilde{\delta}B^2\tau_{s0}\right)z \quad (15)$$

Substituting dependence (15) into equation (13), we obtain that the evolution of the position of the soliton maximum spectrum intensity is also a linear function

$$\omega_{\max}(z) = -\left(\xi - h\cdot\delta_0\right)(\varepsilon_0 - 1)\tilde{\delta}B^2 exp\left(-\tilde{\delta}B^2\tau_{s0}\right)z \quad (16)$$

As it follows from equations (14) and (16), the coefficient at z in equation (16) differ from the coefficient of the second degree of z in equation (16) by the multiplier D.

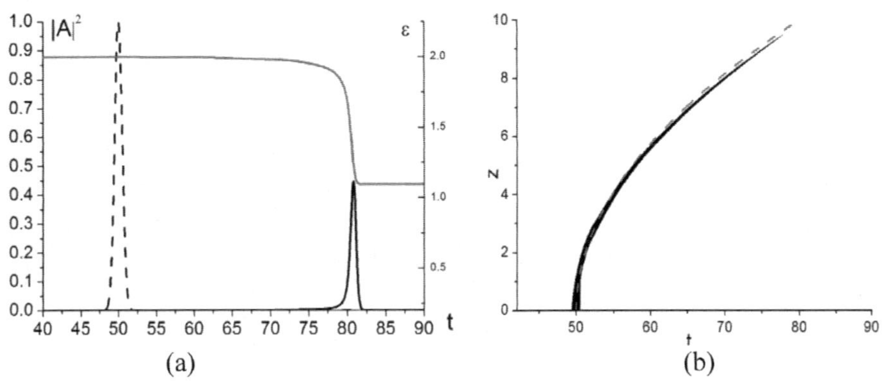

Figure 2. Decelelerating soliton shape (black solid line), incident Gaussian pulse shape (black dashed line) and aspect ratio distribution (red line) at the section $z = 10$ (a), maximal intensity position (b) at the incident Gaussian pulse propagation under the one-photon absorption in the medium with negative phase grating. Dashed line in the figure b is given by (17).

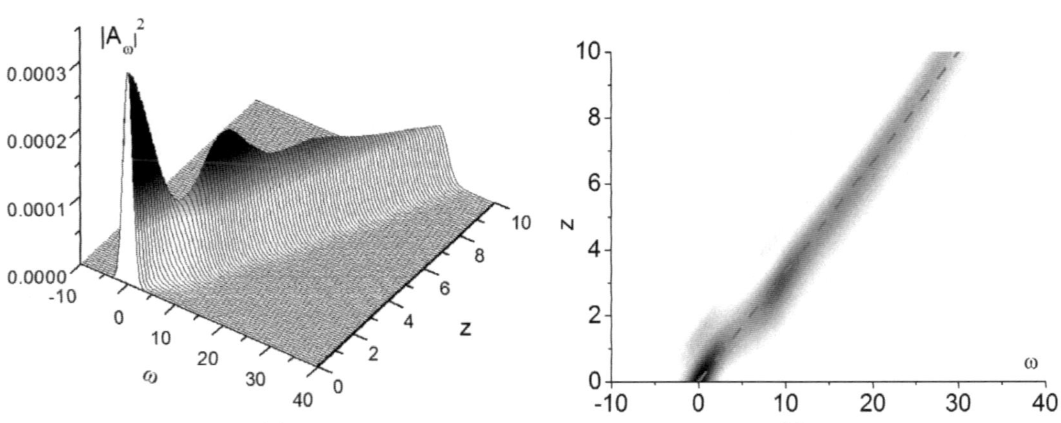

Figure 3. Decelerating soliton spectrum evolution (a), maximum spectrum intensity position (b) at the incident Gaussian pulse propagation under the one-photon absorption in the medium with negative phase grating. Dashed line in the figure b is given by (18).

For the decelerating soliton under consideration, the following values of approximation soliton (6) parameters follow from computer simulation results [22]

$$b_0 = 0,\ \ \tau_{s0} = 0.5,\ \ B_0 \approx 0.668$$

So, the dependences (14) and (16) take the following form

$$t_c(z) = 50 + 0.3z^2 \tag{17}$$

$$\omega_{\max}(z) \approx 3z\,. \tag{18}$$

Dependence (17) was obtained in [22] and is shown in figure 2b by the red dashed line. Dependence (18) is depicted in figure 3b by the dashed line. Good coincidence with the computer simulation results is well seen in figure 3b.

Figure 3 shows the spectrum evolution for the decelerating soliton. As it is well seen, the soliton spectrum undergoes blue shifting – during the soliton propagation, its spectrum preserves its shape and shifts into the area of higher (slower) frequencies (figure 3a).

4.2. Acceleration soliton

In paper [21,23], we have shown that the accelerating soliton is formed in the medium containing nanorods under two-photon absorption and positive phase grating for the following parameters $\delta_0 = 0.005$, $\xi = 5$, $\tilde{\delta} = 5$, $D = 0.1$ (figure 4).

The soliton shape of the left sub-pulse is confirmed by its spectrum evolution (figure 5): the left part of the spectrum gradually shifts into the area of lower frequencies (red shifting).

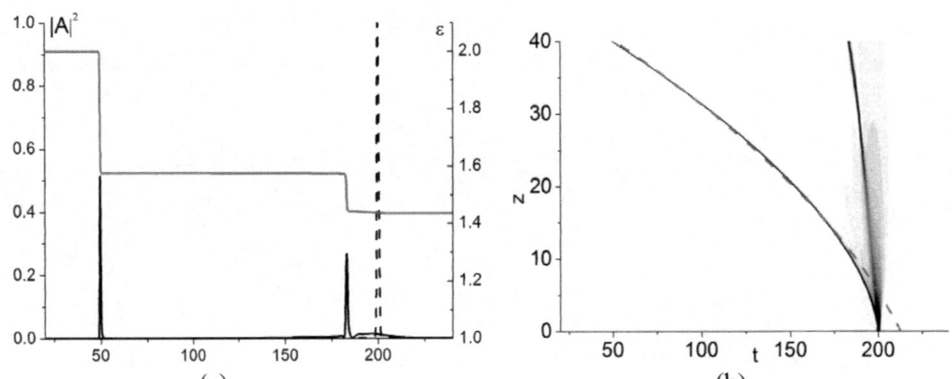

Figure 4. Accelerating sub-pulses shapes (black solid line) and aspect ratio distribution (red line) at the section $z = 40$, incident Gaussian pulse shape (black dashed line) (a), maximal intensity position (b) at the incident Gaussian pulse propagation under the two-photon absorption in the medium with positive phase grating. Dashed line in the figure b is given by (19).

The left sub-pulse acceleration can be described by the following quadratic dependence obtained in [23] (red dashed line in figure 4):

$$t_c(z) = 215 - 3z - 0.05z^2 \tag{19}$$

The absolute value of coefficient at the second degree of z in dependence (19), is about five times smaller than the corresponding coefficient in our analytical dependence for the soliton centre evolution [23]

$$t_c(z) = t_0 + 2\frac{D}{\tau_{s0}} b_0 z - D\xi\tilde{\delta}(\varepsilon_0 - 1)B_0^{\,6}\; exp\left(-2/3\,\tilde{\delta}B_0^{\,4}\tau_{s0}\right)z^2 \;, \tag{20}$$

calculated for approximation soliton parameters obtained in computer experiment for the left sub-pulse at section z=40:

$$\tau_{s0} = 0.3\,,\;\; B_0 \approx 0.748\,. \tag{21}$$

Dependence (20) takes place if the soliton amplitude remains the same for a large trace of propagation which is valid for the parameters under consideration. Under this assumption, the evolution of the soliton frequency along the z-coordinate is a linear function [23]

$$b(z) = b_0 - \xi(\varepsilon_0 - 1)\tilde{\delta}\tau_{s0}B_0^{\,6}exp\left(-2/3\,\tilde{\delta}B_0^{\,4}\tau_{s0}\right)z \tag{22}$$

Substituting (22) into (13), one obtains the linear evolution of the maximum intensity position of the soliton spectrum along the z-coordinate

$$\omega_{\max}(z) = -\xi(\varepsilon_0 - 1)\tilde{\delta}B_0^{\,6}exp\left(-2/3\,\tilde{\delta}B_0^{\,4}\tau_{s0}\right)z \tag{23}$$

Taking into account parameters (21) of soliton approximation (6), it is possible to calculate the coefficient at z in (23), which absolute value occurs about five times bigger than the corresponding coefficient in the linear approximation

$$\omega_{\max}(z) = -0.66z \tag{24}$$

shown in figure 9b by the red dashed line.

As it follows from equations (20) and (22), the coefficient at z in formula (20) differ from the coefficient of the second degree of z in equation (22) by the multiplier D. So, it is possible to use equation (19) to calculate the coefficient at z in dependence (23):

$$\omega_{\max}(z) = -0.5z$$

which is close to approximation (24).

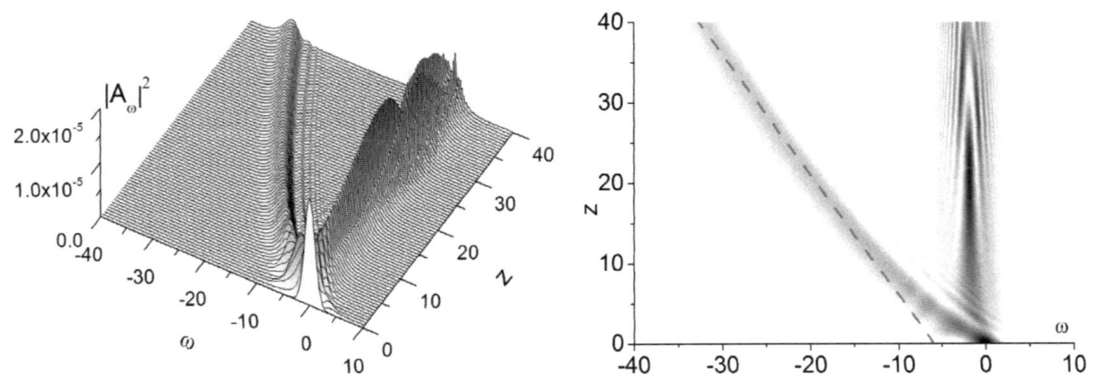

(a) (b)

Figure 5. Accelerating soliton spectrum formation (a), maximal spectrum intensity position (b) at the incident Gaussian pulse propagation under the two-photon absorption in the medium with positive phase grating. Dashed line in the figure b is given by (24).

5. Conclusions

We investigated numerically and analytically the spectrum evolution of the nonlinear chiped solitons - accelerating and slowing down solitons, which are formed at the laser radiation propagation in the medium containing nanorods. We considered decelerating soliton which is formed under one-photon absorption and accelerating soliton which is formed under two-photon absorption. Earlier we have shown that the spectrum of the accelerating soliton undergoes red shifting, while the spectrum of the slowing down soliton undergoes blue shifting.

Using the approximate analytical solution for accelerating and slowing down soliton, we derive the approximate analytical law for the spectrum maximum intensity evolution. We confirmed this law by the computer simulation results. It is essential that the spectrum maximum intensity evolution obeys a linear law with the negative or positive slope depending on the soliton type.

Acknowledgments
The investigation was made using support of the Russian Science Foundation (Grant № 14-21-00081).

References

[1] Zijlstra P, Chon J W M, and Gu M 2009 Five-dimensional optical recording mediated by surface plasmons in gold nanorods *Nature* **459** 410-413
[2] Ditlbacher H, Krenn J R, Lamprecht B, Leitner A, and Aussenegg F R 2000 Spectrally coded optical data storage by metal nanoparticles *Opt. Lett.* **25** 563–565
[3] Wilson O, Wilson G J, and Mulvaney P 2002 Laser writing in polarized silver nanorod films *Adv. Mater.* **14** 1000–1004
[4] Pérez-Juste J, Rodríguez-González B, Mulvaney P, and Liz-Marzán L M 2005 Optical control and patterning of gold-nanorod-poly(vinyl alcohol) nanocomposite films *Adv. Funct. Mater.* **15** 1065–1071
[5] Taylor A B, Kim J and Chon J W M 2012 Detuned surface plasmon resonance scattering of gold nanorods for continuous wave multilayered optical recording and readout *Opt. Express* **20** (5) 5069–5081
[6] Roxworthy B J, Bhuiya A M, Inavalli V V G K, Chen H and Toussaint K C, Jr. 2014 Multifunctional plasmonic film for recording near-field optical intensity *Nano Let.* **14** 4687-4693
[7] Del Fatti N, Christofilos D and Vallée F 2008 Optical response of a single gold nanoparticle *Gold Bulletin* **41**(2) 147-157
[8] Wijers C M J, Chu J-H, Liu J L and Voskoboynikov O 2006 Optical response of layers of embedded semiconductor quantum dots *Phys. Rev. B* **74** 035323
[9] Mohammadi A, Sandoghdar V and Agio M 2008 Gold nanorods and nanospheroids for enhancing spontaneous emission *New Journal of Physics* **10** 105015
[10] Jin R, Jureller J E, Kim H Y and Scherer N F 200 5Correlating second harmonic optical responses of single Ag nanoparticles with morphology *J. Am. Chem. Soc.* **127** 12482-12483
[11] Drachev V P, Buin A K, Nakotte H and Shalaev V M 2004 Size dependent χ^3 for conduction electrons in Ag nanoparticles *Nano Lett.* **4** (8) 1535-1539
[12] Eustis S and El-Sayed M A 2006 Why gold nanoparticles are more precious than pretty gold: Noble metal surface plasmon resonance and its enhancement of the radiative and nonradiative properties of nanocrystals of different shapes *Chemical Society Reviews* **35**, 209-217
[13] Maldonado M, Baltar H T M C M., Gomes A S L, Vaia R, Park K, Che J, Hsiao M, de Araújo C B, Baev A and Prasad P N 2017 Coupled-plasmon induced optical nonlinearities in anisotropic arrays of gold nanorod clusters supported in a polymeric film *J. Applied Physics* **121**, 143103.
[14] Reyna A S and de Araújo C B 2014 Nonlinearity management of photonic composites and

observation of spatial-modulation instability due to quintic nonlinearity *Phys. Rev. A* **89**, 063803

[15] Taylor A B, Siddiquee A M and Chon J W M 2014 Below melting point photothermal reshaping of single gold nanorods driven by surface diffusion *ACS Nano* **8** 12071-12079

[16] Gan Y and Jiang S 2013 Ultrafast Laser-Induced Premelting and Structural Transformation of Gold Nanorod *J. Appl. Phys*. **113** 073507

[17] Wang N, Rokhlin S I and Farson D F 2011 Ultrafast laser melting of Au nanoparticles: Atomistic simulations *J Nanopart Res* 1**3** (10), 4491

[18] Sakai H 1996 Surface-Induced Melting of Small Particles *Surf. Sci.* **351**, 285–291

[19] Goswami G K and Nanda K K 2010 Size-Dependent Melting of Finite-Length Nanowires *J. Phys. Chem. C* **114** (34), 14327–14331

[20] Trofimov VA and Lysak T M 2016 Self-similar mode of laser pulse propagation in a medium containing nanorods *J. Opt.* **18** (1) 025501

[21] Trofimov V A and Lysak T M 2016 Acceleration of light and soliton formation due to nonlinear absorption of femtosecond laser pulse energy by the medium containing nanorods *J. Opt. Soc. Am. B* **33** (1) 62-74

[22] Trofimov V A, Lysak T M 2017 Decelerating chirped soliton formation at femtosecond laser pulse propagation in a medium with one photon absorption and gold nanorods *J. Opt.* **19** (1) 015003

[23] Trofimov V A, Lysak T M 2017 Superluminality phenomenon at a femtosecond laser pulse propagation in a medium containing nanorods *J. of Nanophotonics* **11** (2) 026003

[24] Trofimov V A, Lysak T M 2016 Soliton formarion and the effects of superluminality due to nonlinear absorption of femtosecond laser pulse energy by the medium containing nanorods *J. Phys.: Conf. Ser.* **691** 012012

[25] Trofimov V A, Lysak T M and Lan Sh 2012 Influence of ellipticity of nanorods on both TPA of femtosecond laser pulse and transformation of pulse spectrum *Proc. of SPIE* **8424** 842435

[26] Boyd G T, Yu Z H and Shen Y R 1986 Photoinduced Luminescence From the Noble-Metals and Its Enhancement On Roughened Surfaces *Phys. Rev. B* **33**, (12) 7923-7936

[27] Link S, Mohamed M B and El-Sayed M A 1999 Simulation of the optical absorption spectra of gold nanorods as a function of their aspect ratio and the effect of the medium dielectric constant *J. Phys. Chem. B* **103** (16) 3073-3077

Theoretical investigation of the hyper-Raman scattering in hexagonal semiconductors under two-photon excitation near resonance with the $A_{n=2}$ exciton level

L E Semenova

Prokhorov General Physics Institute, Russian Academy of Sciences, Vavilov St. 38, 119991, Moscow, Russia

E-mail: sl@kapella.gpi.ru

Abstract. The hyper-Raman scattering of light by LO-phonons under two-photon excitation near resonance with the $A_{n=2}$ exciton level in the wurtzite semiconductors A^2B^6 was theoretically investigated, taking into account the influence of the complex structure of the top valence band.

1. Introduction

The experimental investigations of the hyper-Raman scattering (HRS) of light in the conditions of proximity of the doubled energy of incident photons to the band gap E_{cv} in the semiconductors A^2B^6 were reported in a number of publications [1-4]. The HRS features observed were explained within the framework of the excitonic model where the hydrogen-like Wannier excitons are considered as intermediate virtual states of an electronic system [5-6]. In accordance with this model the main contribution to the HRS of light by LO-phonons is given by the sequence of intermediate virtual excitonic states, *s-p-s*. The hyper-Raman process appropriate to it is induced by the intraband Frohlich exciton-lattice interaction and is described within the framework of the two-band model. And its contribution depends on the scattering geometry.

The goal of this work is theoretical investigation of the resonant HRS of light by LO-phonons in the hexagonal crystals A^2B^6 in which the top valence band is split into three closely-lying subbands [7-8]. Consequently, the dipole transitions between the excitons related to the same subband as well as the different ones can participate in the hyper-Raman process. Besides, of special interest is the HRS under two-photon excitation near the $A_{n=2}$ level if the influence of the complex structure of the top valence band on it is taken into account by the use of the perturbation theory.

2. Basic formulas

When the doubled frequency of exciting radiation is close to the fundamental absorption edge of a semiconductor, the hyper-Raman scattering is described by the following sequence of processes: absorption of two incident photons with the wave vector $\boldsymbol{q}_L$, energy $\hbar\omega_L$ and polarization $\boldsymbol{\varepsilon}_L$, a creation of a phonon with the wave vector $\boldsymbol{q}$ and energy $\hbar\omega_P$ and an emission of a photon of scattered light ($\boldsymbol{q}_S$, $\hbar\omega_S$, $\boldsymbol{\varepsilon}_S$) [5,6]. In the initial and final states an electron system of a crystal is assumed to be in the ground state $|\Psi_0\rangle$, but its intermediate virtual states are the Wannier excitons formed by the transition of an electron from the degenerate (or almost degenerate) valence band v to the excitonic level $E_\lambda(\boldsymbol{K})$ near the conduction band c [9]

$$\left|\Psi_\lambda(\boldsymbol{K})\right\rangle = \sum_{j_e=1}^{s_e}\sum_{j_h=1}^{s_h}\int d\boldsymbol{k} A_{j_e j_h}^{(\lambda)}(\boldsymbol{k};\boldsymbol{K})\hat{\psi}_{j_e}^{+}(\mathbf{k})\hat{\psi}_{j_h}(\boldsymbol{k}-\boldsymbol{K})\left|\Psi_0\right\rangle \quad (1)$$

where $\boldsymbol{K}$ is the exciton wave vector, $\hat{\psi}_n^{+}(\boldsymbol{k})$ and $\hat{\psi}_n(\boldsymbol{k})$ are the operators of creation and annihilation of an electron with the wave vector $\boldsymbol{k}$ in the n band, the indices j_h and j_e denote the states in the valence band v and the conduction band c, s_v and s_c are their degeneracy multiplicities.

Because the wave vectors of the photons and the phonons ($\boldsymbol{q}_P = 2\boldsymbol{q}_L - \boldsymbol{q}_S$), which participate in the HRS, are small, the scattering process is assumed to occur in the Brillouin zone center (K~0).

The cross section of the resonant hyper-Raman scattering can be written as [6]

$$\frac{d\sigma}{d\Omega} = \frac{2\pi\hbar e^6 n_S \omega_S^2 (N_P+1) V N_L}{m^6 c^4 \omega_L^3 n_L^3}|B|^2 \quad (2)$$

where e and m are the charge and the mass of an electron, c is the light speed, $n_{L(S)}$ is the refractive index for the frequency $\omega_{L(S)}$, V is the crystal volume, N_L is the density of exciting radiation, N_P is the a number of phonons, B has a form

$$B = \varepsilon_\alpha^{(L)}\varepsilon_\beta^{(L)}\varepsilon_\gamma^{(S)} \sum_{\Lambda_1,\Lambda_2,\Lambda_3} \frac{\Pi_{0\Lambda_3}^{\gamma} P_{\Lambda_3\Lambda_2} \Pi_{\Lambda_2\Lambda_1}^{\beta} \Pi_{\Lambda_1 0}^{\alpha}}{\left(E_{\Lambda_3}-\hbar\omega_S\right)\left(E_{\Lambda_2}-2\hbar\omega_L\right)\left(E_{\Lambda_1}-\hbar\omega_L\right)}. \quad (3)$$

Here the sum runs over all intermediate virtual exciton states $\Lambda = (c, v, \lambda)$.

In this paper the hyper-Raman scattering of light is considered within the framework of the two-band model, and only the dipole-allowed transitions between excitonic states are taken into account. Because of this, the matrix elements of the exciton-photon interaction are described as [9-11]

$$\Pi_{\Lambda 0}^{\alpha} = \sum_{j_e, j_h} \pi_{j_e j_h}^{\alpha} \Phi_{j_e j_h}^{(\lambda)*}(0) \quad (4)$$

and

$$\Pi_{\Lambda'\Lambda}^{\beta} = \sum_{j_e'', j_e', j_h'', j_h'} \left[M_{j_e'' j_e'}^{\beta'\beta}\delta_{j_h'' j_h'} - M_{j_h' j_h''}^{\beta'\beta}\delta_{j_e'' j_e'}\right]\left\langle \Phi_{j_e'' j_h''}^{(\lambda'')}\right|\hat{p}_{\beta'}\left|\Phi_{j_e' j_h'}^{(\lambda')}\right\rangle \quad (5)$$

where $\hat{p}_\alpha = -i\hbar\, d/dr_\alpha$, $\boldsymbol{\pi}_{cv}$ is the interband matrix element of the momentum operator, $\Phi_{j_e j_h}^{(\lambda)}(\boldsymbol{r})$ is the wave function of the relative electron-hole motion. The parameter $M_{jj'}^{\alpha\beta}$ is determined by [9,12]

$$M_{jj'}^{\alpha\beta} = \frac{1}{m}\sum_i{}' \left[\frac{\pi_{ji}^{\alpha}\pi_{ij'}^{\beta}}{E_{ji}} + \frac{\pi_{ji}^{\beta}\pi_{ij'}^{\alpha}}{E_{j'i}}\right]. \quad (6)$$

In the equation (3) the matrix element of the intraband Frohlich exciton–lattice interaction $P_{\Lambda'\Lambda}$ is considered in the approximation of the zero wave vector of the phonon [13].

The wave functions of the relative motion of an electron and a hole $\Phi_{j_e j_h}^{(\lambda)}(\boldsymbol{r})$ are assumed to satisfy a system of the effective mass approximation equations [9,14-15].

As may be seen from the equation (5), the matrix element can describe the dipole transitions between the excitons related to the same and different series.

3. Exciton wave functions

With allowance made for anisotropy and the complex valence band of hexagonal semiconductors, the wave functions and the energies in the first approximation for the $A_{n=1}$ and $B_{n=1}$ exciton levels are reported in Ref. [16] to be obtained. In this paper the analogous approach is applied to find the first order corrections to the wave functions and the energy of the $A_{n=2}$ level. In the case of the wurtzite semiconductors A^2B^6 the set of the equations for the wave functions $\Phi^{(\lambda)}_{j_e j_h}(\mathbf{r})$ can be written as [16]

$$\left\{\frac{1}{2\mu^{\perp}_{j_e j_h}}\left(\hat{p}_x^2+\hat{p}_y^2\right)+\frac{1}{2\mu^{\parallel}_{j_e j_h}}\hat{p}_z^2+U_{eh}(\mathbf{r})\right\}\Phi^{(\lambda)}_{j_e j_h}(\mathbf{r})-\sum_{j'_h\neq j_h}D^{\beta\alpha}_{j'_h j_h}\hat{p}_\alpha\hat{p}_\beta\Phi^{(\lambda)}_{j_e j'_h}(\mathbf{r})=\left(E_\lambda-\varepsilon_{j_e j_h}\right)\Phi^{(\lambda)}_{j_e j_h}(\mathbf{r}) \quad (7)$$

where $E_\lambda = E_\lambda(0)$, $\varepsilon_{j_e j_h}$ is the energy gap in the Brillouin zone center, $\mu_{j_e j_h}$ is the excitonic effective mass, $U_{eh}(\mathbf{r})$ is the Coulomb energy of the electron-hole interaction, $D^{\beta\alpha}_{j'j}$ are the effective mass parameters [9]

$$D^{\alpha\beta}_{jj'}=\frac{1}{2m}\delta_{jj'}\delta_{\alpha\beta}+\frac{1}{m^2}\sum_n\frac{\pi^\alpha_{jn}\pi^\beta_{nj'}}{\varepsilon_{jn}}. \quad (8)$$

Based on the band structure of the considered crystals and the selection rules the nonzero parameters $D^{\beta\alpha}_{j'_h j_h}$ can be found.

Introduction of the some scalar effective mass of the exciton $\overline{\mu}_{j_e j_h}$, as well as in Ref. [16], allows one to write the set of the equations (7) as

$$\left\{\frac{1}{2\overline{\mu}_{j_e j_h}}\left(\hat{p}_x^2+\hat{p}_y^2+\hat{p}_z^2\right)+U_{eh}(\mathbf{r})\right\}\Phi^{(\lambda)}_{j_e j_h}(\mathbf{r})+\hat{G}_{j_e j_h}\Phi^{(\lambda)}_{j_e j_h}(\mathbf{r})-\sum_{j'_h\neq j_h}D^{\beta\alpha}_{j'_h j_h}\hat{p}_\alpha\hat{p}_\beta\Phi^{(\lambda)}_{j_e j'_h}(\mathbf{r})=\left(E_\lambda-\varepsilon_{j_e j_h}\right)\Phi^{(\lambda)}_{j_e j_h}(\mathbf{r}) \quad (9)$$

where

$$\hat{G}_{j_e j_h}=\frac{1}{2}\left\{\left(\frac{1}{\mu^{\perp}_{j_e j_h}}-\frac{1}{\overline{\mu}_{j_e j_h}}\right)\left(\hat{p}_x^2+\hat{p}_y^2\right)+\left(\frac{1}{\mu^{\parallel}_{j_e j_h}}-\frac{1}{\overline{\mu}_{j_e j_h}}\right)\hat{p}_z^2\right\}. \quad (10)$$

It is obvious that the set of the equations (9) is reduced to the individual well-known hydrogen-like equations if the second and third terms are neglected and the anisotropy effect on $U_{eh}(\mathbf{r})$ is ignored. In order to allow for the influence of the complex valence band and anisotropy of the excitonic effective mass, the disregarded terms can be taken into account with the use of the perturbation theory [17]. For this purpose the wave functions $\Phi^{(\lambda)}_{j_e j_h}(\mathbf{r})$ are expanded over the hydrogen-like wave functions $\chi^{(\xi\ell m)}_{j_e j_h}(\mathbf{r})$ [17]

$$\Phi^{(\lambda)}_{j_e j_h}(\mathbf{r})=\sum_{\xi,\ell,m}c^{(\xi\ell m)}_{j_e j_h}\chi^{(\xi\ell m)}_{j_e j_h}(\mathbf{r}). \quad (11)$$

where $(\xi\ell m)$ is the set of quantum numbers characterizing the relative electron-hole motion. After the substitution of the expansion (11), each from the equations of the set (9) is multiplied by $\chi^{(\xi'\ell'm')}_{j_e j_h}(\mathbf{r})^*$ and then the integration over $\mathbf{r}$ is carried out. Thus, the system of the equations for coefficients $c^{(\xi\ell m)}_{j_e j_h}$ is obtained. These coefficients are considered here as $c^{(\xi\ell m)}_{j_e j_h}=a^{(\xi\ell m)}_{j_e j_h}+b^{(\xi\ell m)}_{j_e j_h}$ where $a^{(\xi\ell m)}_{j_e j_h}$ and $b^{(\xi\ell m)}_{j_e j_h}$ correspond to the zero and first approximation. In order to find the first order corrections to the $A_{n=2}$ energy level the coefficients $a^{(\xi\ell m)}_{j_e j_h}$ related to the B and C exciton series are taken to be equal to zero,

IOP Conf. Series: Journal of Physics: Conf. Series **999** (2018) 012012 doi:10.1088/1742-6596/999/1/012012

but in the case of the A series $a_{j_e j_h}^{(\xi \ell m)} = 0$ if $\xi \neq 2$. If ξ' is assumed to be equal to 2 and the second approximation terms are discarded in the obtained system of equations, then the first order corrections to the $A_{n=2}$ level can be derived. In this case the splitting into three sublevels is found. If the discarded terms associated this level with others are taken into account, the $A_{n=2}$ level splits into four sublevels, for which the wave functions in the first approximation were derived. Because the obtained expressions are too cumbersome, they are not present in this paper.

4. Discussion

Let us consider the resonant HRS by LO-phonons in the wurtzite semiconductors on the assumption that the hydrogen-like Wannier excitons participate in the scattering process. The polarization of the incident radiation is supposed to be perpendicular to the axis *z*. If the dipole transitions between the subbands are ignored, then the contributions of the A, B and C series to the HRS cross section depends on the scattering geometry as $B \propto \left(\varepsilon_{\breve{O}}^{(L)}\varepsilon_{\breve{O}}^{(S)} + \varepsilon_{Y}^{(L)}\varepsilon_{Y}^{(S)}\right)\left(\varepsilon_{X}^{(L)}\hat{q}_X + \varepsilon_{Y}^{(L)}\hat{q}_Y\right)$ where $\hat{\boldsymbol{q}}$ is the unit vector in the direction of the phonon wave vector.

As may be seen from the equations (6) and (8), the parameters $M_{\nu\nu'}^{\alpha\beta}$ ($\nu' \neq \nu$) can be associated with $D_{\nu\nu'}^{\alpha\beta}$ by the correlation $M_{\nu\nu'}^{\alpha\beta} = m\left(D_{\nu\nu'}^{\alpha\beta} + D_{\nu'\nu}^{\alpha\beta *}\right)$. Consequently, if the nonzero parameters $D_{\nu\nu'}^{\alpha\beta}$ for the wurtzite crystals are known, analysis of the contributions of the transitions between the excitons of the different series to the HRS can be performed. Because of this, of some interest are the scattering geometries *z(xx,y')x'* and *z(xx,z)x'* when the incident radiation with the polarized $\boldsymbol{\varepsilon}_{\mathrm{L}} \parallel x$ is traveling along the *z* axis and the scattered light propagates normally on *z* and at the angle $\phi_{\mathrm{L}} = \pi/4$ to the *x* axis. In this case the sequences of the exciton states related to the same series participate in the hyper-Raman process if $\boldsymbol{\varepsilon}_{\mathrm{S}} \perp z$. But due to the inclusion of the dipole transitions between the excitons of different series the considered HRS is possible when $\boldsymbol{\varepsilon}_{\mathrm{S}} \parallel z$. In this case the pertinent hyper-Raman process is described as the two-photon transition to the *p*-exciton state of the B or C series and the indirect transition to the ground state. And its contribution grows as the doubled frequency of exciting radiation is approached to resonance with the $B_{n=2}$ exciton level. Besides, it was found that allowance made for the transitions between the *s*- and *p*-excitons related to the different subbands leads to the hyper-Raman scattering for the scattering geometry *y(xx,y)z*.

The frequency dependences of the HRS cross section dσ/dΩ for CdS and CdSe were obtained. The results are present in Fig. 1 where the curves (1) and (2) illustrate dσ/dΩ calculated for the scattering geometry *z(xx,y')x'* as a function of the doubled energy of incident photons if the transitions between the excitons associated with the different subbands are ignored (1) and are taken into account (2). The curves (3) and (4) show the cross section computed for the geometries *z(xx,z)x'* and *y(xx,y)z*, respectively. Calculations were performed for the following parameters of semiconductors: $R_{\mathrm{A}} = R_{\mathrm{B}} = 0.028\,eV$, $R_{\mathrm{C}} = 0.026\,eV$ [7], $E_{cv}^{\mathrm{A}} = 2.578\,eV$,V $E_{cv}^{\mathrm{B}} = 2.593\,eV$, $E_{cv}^{\mathrm{C}} = 2.656\,eV$ [18], $\hbar\omega_{\mathrm{P}} = 0.038 eV$ [19], $\Gamma_{\mathrm{A}} = 0.004\,eV$, $\Gamma_{\mathrm{B}} = 0.005\,eV$, and $\Gamma_{\mathrm{C}} = 0.026\,eV$ [8] for CdS and $R_{\mathrm{A}} = 0.0158\,eV$, $R_{\mathrm{B}} = R_{\mathrm{C}} = 0.0161\,eV$, $E_{cv}^{\mathrm{A}} = 1.833\,eV$, $E_{cv}^{\mathrm{B}} = 1.8571\,eV$, $E_{cv}^{\mathrm{C}} = 2.272\,eV$ [8], $\hbar\omega_{\mathrm{P}} = 0.027\,eV$ [19], $\Gamma_{\mathrm{A}} = 0.004\,eV$, $\Gamma_{\mathrm{B}} = 0.006\,eV$, and $\Gamma_{\mathrm{C}} = 0.080\,eV$ [8] for CdSe where $\Gamma_{\mathrm{A(B,C)}}$ and $R_{\mathrm{A(B,C)}}$ are the lifetime broadening of the exciton states and the Rydberg of the A (B, C) series, E_{cv} is the energy gap. The data on the effective masses of an electron and a hole in CdS were used from Ref. [18]. In the case of CdSe the exciton effective masses are assumed to little differ in a magnitude. Besides, it is supposed that the terms including the transitions between the nearest bands give the main contribution to the parameters $D_{\nu\nu'}^{\alpha\beta}$. Under summation over the intermediate virtual states the Green's function approach was applied [6,20].

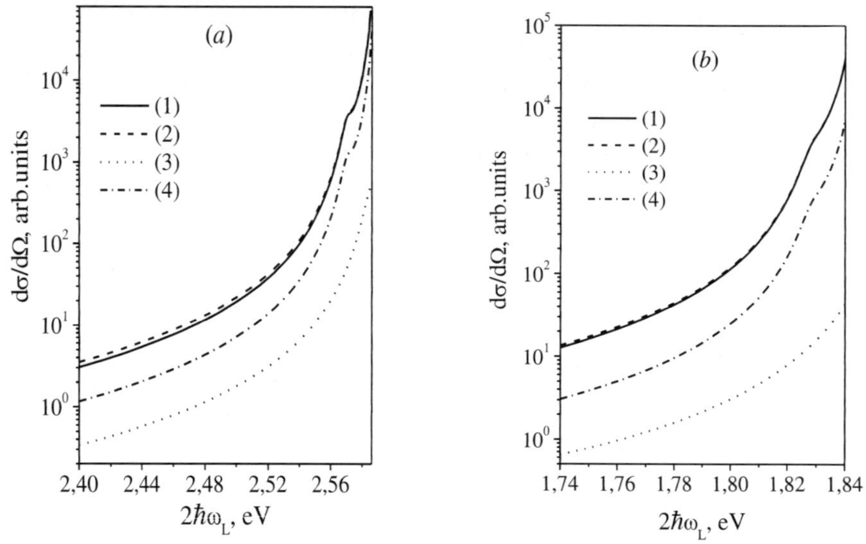

Figure 1. The HRS cross section calculated for CdS (*a*) and CdSe (*b*) as a function of the doubled energy of incident photons.

As can be seen from Fig.1, impossibility of participation of the sequences of the intermediate exciton states related to the same subbands in the HRS leads to the appreciable decrease in the cross section. It has been known that the hyper-Raman scattering induced by the two-photon transitions to the *s*-excitonic states can occur too [2,6,21-22]. In general case the contribution of this mechanism is considerably less than the one of the dominant sequence of the intermediate states [6]. However, under experimental investigation of the HRS in ZnSe on the frequency dependence of the scattering intensity a weak enhancement was reported to be found for the two-photon resonance with the 1s exciton state [2]. Thus, the hyper-Raman processes taking into account the transitions between the subbands along with the scattering mechanisms including the two-photon transitions to the *s*-exciton states can give a contribution to the cross section when the sequence of the intermediate excitonic states *s-p-s* of the same series do not participate in the HRS.

In order to assess the role of the obtained the corrections to the wave functions the HRS induced by the two-photon transition to the $A_{n=2}$ exciton level was considered for the following scattering geometries: *z(xx,y')x'*, *z(xx,z)x'* and *y(xx,y)z*. Rough estimations of the relative contributions of this hyper-Raman process to the cross section ($d\sigma_A/d\Omega$) were executed for the CdS, CdSe and ZnS semiconductors. As an example the results obtained for a CdS crystal are shown in Fig. 2. As mentioned above, for the configuration *z(xx,y')x'* the sequence of the intermediate exciton states related to the A series dominates in the HRS. The inclusion of the corrections to the wave functions of the $A_{n=2}$ level leads to a decrease of the relative contribution of the considered hyper-Raman process to the cross section. The approximate estimations have shown that in this case $d\sigma_A/d\Omega$ decreases approximately twice for $2\hbar\omega_L \sim E_{A_{n=2}}$. For the scattering geometry, *z(xx,z)x'* allowance made for the corrections to the wave functions leads to the added HRS process caused by the two-photon transition to the $A_{n=2}$ level. However, its contribution is negligible in comparison with the one calculated for the geometry *z(xx,y')x'*. The rough estimations have shown that in the case of the configuration *y(xx,y)z* the inclusion of the corrections gives rise to a decrease of the relative contribution of the considered process to the cross section too.

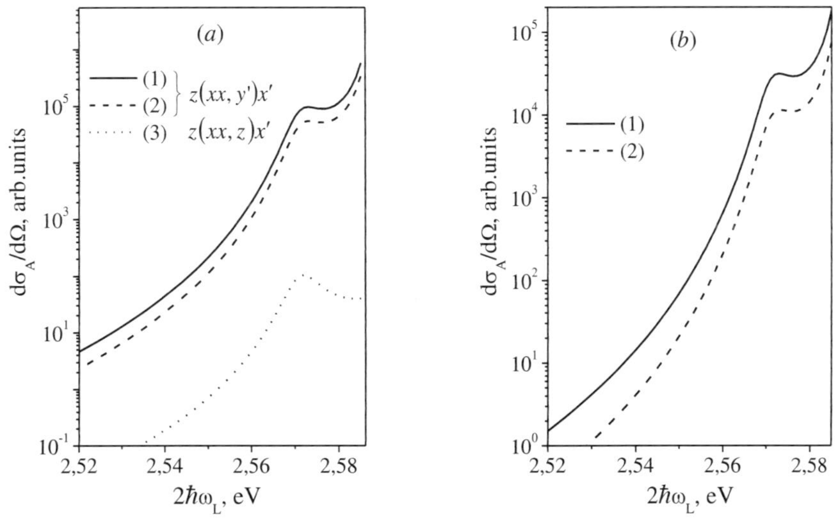

Figure 2. The HRS cross section calculated for the scattering mechanism including the two-photon transition to the $A_{n=2}$ level in a CdS crystal as a function of the doubled energy of incident photons. Calculations were carried out for the configurations *z(xx,y')x', z(xx,z)x'* (*a*) and *y(xx,y)z* (*b*) if the first order corrections to the wave functions are neglected (the curve 1) and are taken into account (the curves 2 and 3).

5. Conclusion

The resonant hyper-Raman scattering by LO-phonons in the wurtzite semiconductors A^2B^6 was treated within the framework two-band model, taking into account the Wannier excitons as the intermediate states of an electronic system. It has been shown that in certain cases the sequences of the intermediate excitonic states, *s-p-s*, can give a contribution to the HRS only due to the inclusion of the transitions between excitons related to the different series.

Taking into account the complex structure of the top valence band and the anisotropy of the exciton effective mass by the perturbation theory, the energies and the wave functions in a first approximation were obtained for the $A_{n=2}$ excitonic level. The hyper-Raman scattering of light by LO-phonons near the two-photon resonance with the $A_{n=2}$ excitonic level was theoretically investigated. Rough estimations obtained for some semiconductors have shown that allowance made for these corrections can exert perceptible influence on the contribution of the hyper-Raman process including the two-photon transition to the $A_{n=2}$ excitonic level to the cross section.

Acknowledgments

This study was partially supported by the Program III.2 of the Russian Academy of Sciences "Fundamental problems of laser technologies".

References

[1] Inoue K, Minami F, Kato Y and Yoshida K 1992 *J. Cryst. Growth* **117** 738-41
[2] Inoue K, Yoshida K, Minami F and Kato Y 1992 *Phys. Rev.* B **45** 8807-10
[3] Matsushita S, Minami F, Yoshida K, Inoue K and Akinaga H 1996 *J. Luminesc.* **66&67** 414-7
[4] Maslov V A, Ondriash K K, Polivanov Yu N, Prokhorov K A, Semenova L E and Chuzavkov Yu L 1996 *Las. Phys.* **6** 132-43
[5] Garcıa-Cristobal A, Cantarero A, Trallero-Giner C and Cardona M 1998 *Phys. Rev.* B **58** 10443-51

[6] Semenova L E and Prokhorov K A 2003 *JETP* **96** 922-31
[7] Reynolds D C 1969 Excitons in II-VI compounds *Optical Properties of Solids* ed S Nudelman and S S Mitra (New York: Plenum Press) chapter 10 pp 239-86
[8] Sobolev V V 1980 *Bands and Excitons of the $A^{II}B^{VI}$ Group Compounds* (Chisinau: Shtiintsa)
[9] Denisov M M and Makarov V P 1973 *Phys. Stat. Sol.* (b) **56** 9-59
[10] Rustagi K C, Pradere F and Mysyrowicz A 1973 *Phys. Rev.* B **8** 2721-32
[11] Mahan G D 1968 *Phys. Rev.* **170** 825-38
[12] Elliott R J 1957 *Phys. Rev.* **108** 1384-9
[13] Ganguly A K and Birman J L 1967 *Phys. Rev.* **162** 806-16
[14] Makarov V P 1971 *Phys. Stat. Sol.* (b) **44** 475-88
[15] Dresselhaus G 1956 *J. Phys. Chem. Solids* **1** 14-22
[16] Semenova L E, Nikolaeva G Yu and K A Prokhorov 2017 *J. Phys.: Conf. Ser.* **826** 012004
[17] Landau L D and Lishitz E M 1989 *Course of Theoretical Physics* vol 3 (Moscow:Nauka)
[18] Nguyen B A, Nguyen V H, Nguyen T T and Nguyen A V 1982 *Phys. Rev.* B **25** 4075-80
[19] 1991 *Physical Quantities: a Handbook* ed I S Grigor'ev and E Z Meilikhov (Moscow: Energoatomizdat)
[20] Hostler L 1964 *J. Math. Phys.* **5** 825-38
[21] Semenova L E and Prokhorov K A 2004 *Las. Phys. Lett.* **1** 253-8
[22] Garcıa-Cristobal A, Cantarero A, Trallero-Giner C and Cardona M 1999 *Physica* B **263-264** 809-12

Solvable Model of a Generic Trapped Mixture of Interacting Bosons: Many-Body and Mean-Field Properties

S Klaiman[1], A I Streltsov[1,2] and O E Alon[3,4]

[1] Theoretical Chemistry, Physical Chemistry Institute, Heidelberg University, Heidelberg, Germany
[2] Institute of Physics, University of Kassel, Kassel, Germany
[3] Department of Mathematics, University of Haifa, Haifa, Israel
[4] Haifa Research Center for Theoretical Physics and Astrophysics, University of Haifa, Haifa, Israel

E-mail: ofir@research.haifa.ac.il

Abstract. A solvable model of a generic trapped bosonic mixture, N_1 bosons of mass m_1 and N_2 bosons of mass m_2 trapped in an harmonic potential of frequency ω and interacting by harmonic inter-particle interactions of strengths λ_1, λ_2, and λ_{12}, is discussed. It has recently been shown for the ground state [J. Phys. A **50**, 295002 (2017)] that in the infinite-particle limit, when the interaction parameters $\lambda_1(N_1-1)$, $\lambda_2(N_2-1)$, $\lambda_{12}N_1$, $\lambda_{12}N_2$ are held fixed, each of the species is 100% condensed and its density per particle as well as the total energy per particle are given by the solution of the coupled Gross-Pitaevskii equations of the mixture. In the present work we investigate properties of the trapped generic mixture at the infinite-particle limit, and find differences between the many-body and mean-field descriptions of the mixture, despite each species being 100%. We compute analytically and analyze, both for the mixture and for each species, the center-of-mass position and momentum variances, their uncertainty product, the angular-momentum variance, as well as the overlap of the exact and Gross-Pitaevskii wavefunctions of the mixture. The results obtained in this work can be considered as a step forward in characterizing how important are many-body effects in a fully condensed trapped bosonic mixture at the infinite-particle limit.

1. Introduction

Ever since the first experimental demonstration of trapped Bose-Einstein condensates (BECs) in ultra-cold quantum gases [1-3], the connection between the microscopic many-particle Hamiltonian and the macroscopic Gross-Pitaevskii, mean-field theory has drawn much attention [4-14]. Whereas properties like the energy and density per particle, and being 100% condensed, are exactly reproduced in the infinite-particle limit by the Gross-Pitaevskii theory [5-8], other properties, like the variance of many-particle operators and, ultimately, the many-particle wavefunction itself generally are not reproduced by the mean-field theory [10-14].

Mixtures of BECs have also attracted a lot of interest in their static, thermal, and out-of-equilibrium properties [15-40], yet the connection between their microscopic many-body Hamiltonian and the mean-field solution for mixtures has only recently received attention [41-46]. In particular, that each of the species in the ground state of a generic trapped mixture is, in the infinite-particle limit, 100% condensed has been recently shown within the exactly-solvable harmonic-interaction model for trapped mixtures [43]. The harmonic-interaction model has widely been studied for indistinguishable bosons [47-55], fermions [53-58], and bosonic mixtures [59-63]. The purpose of the present work is to compute and compare many-body and mean-field properties of the 100% condensed generic mixture [43] at the infinite-particle limit. The ground-state wavefunction can be prescribed analytically at the exact and mean-field levels, which facilitates a transparent study of properties of the mixture at both levels of theory. We extend and broaden previous results obtained in the specific case of a symmetric mixture [42]. We concentrate on the center-of-mass position and momentum variances, their uncertainty product, and angular-momentum variance of the whole mixture as well as of each species in the mixture. Generally, in comparison with the textbook single-particle case [64] the variances and uncertainty product of many-particle operators are more involved [10, 11, 13], see in this context also [65, 66]. Finally, motivated by recent results in the single-species case [12, 14], we evaluate explicitly the overlap of the exact and Gross-Piatesvkii wavefunctions of the generic mixture. A common line of our investigation is how intra-species and inter-species interaction parameters (in combination with the masses) influence the respective properties of the mixture at the infinite-particle limit.

2. The harmonic-interaction model for a generic trapped mixture

We start from the many-particle Hamiltonian ($\hbar = 1$)

$$\hat{H}(\mathbf{x}_1,\ldots,\mathbf{x}_{N_1},\mathbf{y}_1,\ldots,\mathbf{y}_{N_2}) = \\ = \sum_{i=1}^{N_1}\left(-\frac{1}{2m_1}\frac{\partial^2}{\partial\mathbf{x}_i^2}+\frac{1}{2}m_1\omega^2\mathbf{x}_i^2\right)+\sum_{j=1}^{N_2}\left(-\frac{1}{2m_2}\frac{\partial^2}{\partial\mathbf{y}_j^2}+\frac{1}{2}m_2\omega^2\mathbf{y}_j^2\right)+ \\ +\lambda_1\sum_{1\le i<l}^{N_1}(\mathbf{x}_i-\mathbf{x}_l)^2+\lambda_2\sum_{1\le j<m}^{N_2}(\mathbf{y}_j-\mathbf{y}_m)^2+\lambda_{12}\sum_{i=1}^{N_1}\sum_{j=1}^{N_2}(\mathbf{x}_i-\mathbf{y}_j)^2 \qquad (1)$$

which describes a generic trapped mixture of N_1 bosons of mass m_1 (species 1) and N_2 bosons of mass m_2 (species 2) trapped in an harmonic potential of frequency ω and interacting by harmonic intra-species and inter-species interactions of strengths λ_1, λ_2, and λ_{12}. Positive values of λ_1, λ_2, and λ_{12} mean attraction, and negative values repulsion. Of course, a combination of repulsive and attractive intra-species and inter-species interactions is possible [43]. The Hamiltonian (1) can be diagonalized by moving from the laboratory frame (i.e., transforming the Cartesian coordinates) to a set of Jacobi coordinates

$$\mathbf{Q}_s = \frac{1}{\sqrt{s(s+1)}}\sum_{i=1}^{s}(\mathbf{x}_{s+1}-\mathbf{x}_i), \qquad 1\le s\le N_1-1,$$

$$\mathbf{Q}_{N_1-1+s} = \frac{1}{\sqrt{s(s+1)}}\sum_{j=1}^{s}(\mathbf{y}_{s+1}-\mathbf{y}_j), \qquad 1\le s\le N_2-1,$$

$$\mathbf{Q}_{N-1} = \sqrt{\frac{N_2}{N_1}}\sum_{i=1}^{N_1}\mathbf{x}_i-\sqrt{\frac{N_1}{N_2}}\sum_{j=1}^{N_2}\mathbf{y}_j, \quad \mathbf{Q}_N = \frac{m_1}{M}\sum_{i=1}^{N_1}\mathbf{x}_i+\frac{m_2}{M}\sum_{j=1}^{N_2}\mathbf{y}_j. \qquad (2)$$

Note that the Jacobian of the transformation satisfies $\left|\frac{\partial(\mathbf{Q}_1,\ldots,\mathbf{Q}_N)}{\partial(\mathbf{x}_1,\ldots,\mathbf{x}_{N_1},\mathbf{y}_1,\ldots,\mathbf{y}_{N_2})}\right| = 1$.

The transformed many-particle Hamiltonian is separable and reads

$$\hat{H}(\mathbf{Q}_1,\ldots,\mathbf{Q}_N) = \sum_{s=1}^{N_1-1}\left(-\frac{1}{2m_1}\frac{\partial^2}{\partial \mathbf{Q}_s^2}+\frac{1}{2}m_1\Omega_1^2\mathbf{Q}_s^2\right)+\sum_{s=N_1}^{N-2}\left(-\frac{1}{2m_2}\frac{\partial^2}{\partial \mathbf{Q}_s^2}+\frac{1}{2}m_2\Omega_2^2\mathbf{Q}_s^2\right)+$$
$$+\left(-\frac{1}{2M_{12}}\frac{\partial^2}{\partial \mathbf{Q}_{N-1}^2}+\frac{1}{2}M_{12}\Omega_{12}^2\mathbf{Q}_{N-1}^2\right)+\left(-\frac{1}{2M}\frac{\partial^2}{\partial \mathbf{Q}_N^2}+\frac{1}{2}M\omega^2\mathbf{Q}_N^2\right), \tag{3}$$

where $M_{12} = \frac{m_1 m_2}{M}$ and $M = N_1 m_1 + N_2 m_2$ are the reduced and total mass, respectively, and

$$\Omega_1 = \sqrt{\omega^2+\frac{2}{m_1}\left(\Lambda_1\frac{N_1}{N_1-1}+\Lambda_{21}\right)}, \qquad \Omega_2 = \sqrt{\omega^2+\frac{2}{m_2}\left(\Lambda_2\frac{N_2}{N_2-1}+\Lambda_{12}\right)},$$
$$\Omega_{12} = \sqrt{\omega^2+2\left(\frac{\Lambda_{12}}{m_2}+\frac{\Lambda_{21}}{m_1}\right)}, \qquad \omega \tag{4}$$

are the eigen-frequencies of the decoupled oscillators. Ω_1 and Ω_2 are associated with the intra-species relative coordinates, Ω_{12} with the inter-species relative coordinate (between the center-of-mass of species 1 and the center-of-mass of species 2), and the eigen-frequency of the center-of-mass coordinate of the whole mixture is equal to the trapping potential ω. Herein, the so called mean-field interaction parameters are defined, $\Lambda_1 = \lambda_1(N_1-1)$, $\Lambda_2 = \lambda_2(N_2-1)$, $\Lambda_{12} = \lambda_{12}N_1$, and $\Lambda_{21} = \lambda_{12}N_2$, and shall have an important role below.

The ground-state wavefunction of the mixture is thus

$$\Psi(\mathbf{Q}_1,\ldots,\mathbf{Q}_N) = \left(\frac{m_1\Omega_1}{\pi}\right)^{\frac{3(N_1-1)}{4}}\left(\frac{m_2\Omega_2}{\pi}\right)^{\frac{3(N_2-1)}{4}}\left(\frac{M_{12}\Omega_{12}}{\pi}\right)^{\frac{3}{4}}\left(\frac{M\omega}{\pi}\right)^{\frac{3}{4}}\times$$
$$\times e^{-\frac{1}{2}\left(m_1\Omega_1\sum_{s=1}^{N_1-1}\mathbf{Q}_s^2+m_2\Omega_2\sum_{s=N_1}^{N-2}\mathbf{Q}_s^2+M_{12}\Omega_{12}\mathbf{Q}_{N-1}^2+M\omega\mathbf{Q}_N^2\right)} \tag{5}$$

and

$$E = \frac{3}{2}\left[(N_1-1)\sqrt{\omega^2+\frac{2}{m_1}\left(\Lambda_1\frac{N_1}{N_1-1}+\Lambda_{21}\right)}+\right.$$
$$\left.+(N_2-1)\sqrt{\omega^2+\frac{2}{m_2}\left(\Lambda_2\frac{N_2}{N_2-1}+\Lambda_{12}\right)}+\sqrt{\omega^2+2\left(\frac{\Lambda_{12}}{m_2}+\frac{\Lambda_{21}}{m_1}\right)}+\omega\right] \tag{6}$$

is the ground-state energy which can be seen as a functional of the numbers and masses of the particles and the interaction parameters.

It is possible to solve the Hamiltonian (1) at the mean-field level [43]. The Gross-Pitaevskii (mean-field) wavefunction is

$$\Phi^{GP}(\mathbf{x}_1,\ldots,\mathbf{x}_{N_1},\mathbf{y}_1,\ldots,\mathbf{y}_{N_2}) =$$
$$= \left(\frac{m_1\Omega_1^{GP}}{\pi}\right)^{\frac{3N_1}{4}}\left(\frac{m_2\Omega_2^{GP}}{\pi}\right)^{\frac{3N_2}{4}} e^{-\frac{1}{2}\left(m_1\Omega_1^{GP}\sum_{i=1}^{N_1}\mathbf{x}_i^2+m_2\Omega_2^{GP}\sum_{j=1}^{N_2}\mathbf{y}_j^2\right)},$$
$$\Omega_1^{GP} = \sqrt{\omega^2+\frac{2}{m_1}(\Lambda_1+\Lambda_{21})}, \quad \Omega_2^{GP} = \sqrt{\omega^2+\frac{2}{m_2}(\Lambda_2+\Lambda_{12})}. \tag{7}$$

It is a double-product wavefunction made of the (interaction-dressed) Gross-Pitaevskii orbitals

$$\phi_1^{GP}(\mathbf{x}) = \left(\frac{m_1\Omega_1^{GP}}{\pi}\right)^{\frac{3}{4}} e^{-\frac{m_1}{2}\Omega_1^{GP}\mathbf{x}^2}, \quad \phi_2^{GP}(\mathbf{y}) = \left(\frac{m_2\Omega_2^{GP}}{\pi}\right)^{\frac{3}{4}} e^{-\frac{m_2}{2}\Omega_2^{GP}\mathbf{y}^2}. \tag{8}$$

IOP Conf. Series: Journal of Physics: Conf. Series **999** (2018) 012013 doi:10.1088/1742-6596/999/1/012013

Ω_1^{GP} and Ω_2^{GP} are the Gross-Pitaevskii frequencies. For $N_1 \gg 1$ and $N_2 \gg 1$ they coincide with the respective many-body eigen-frequencies Ω_1 and Ω_2 (4).

The mean-field energy per particle is given by

$$\varepsilon^{GP} = \frac{E^{GP}}{N} = \frac{3}{2(\Lambda_{12}+\Lambda_{21})}\left[\Lambda_{12}\sqrt{\omega^2+\frac{2}{m_1}(\Lambda_1+\Lambda_{21})}+\Lambda_{21}\sqrt{\omega^2+\frac{2}{m_2}(\Lambda_2+\Lambda_{12})}\right] \tag{9}$$

and expressed solely via the masses of the bosons and the interaction parameters for any number of particles (N_1 and N_2 and therefore) N. A connection between masses and interaction parameters will further serve below.

3. Properties at the infinite-particle limit

In the two-species infinite-particle limit, i.e., when $N_1 \to \infty$ and $N_2 \to \infty$ such that the interaction parameters Λ_1, Λ_2, Λ_{12}, and Λ_{21} (and therefore $\frac{N_1}{N_2} = \frac{\Lambda_{12}}{\Lambda_{21}}$) are held fixed (hereafter briefly, the infinite-particle limit), we have [43] for the frequencies $\lim_{N\to\infty}\Omega_1 = \Omega_1^{GP}$ and $\lim_{N\to\infty}\Omega_2 = \Omega_2^{GP}$, for the energy per particle

$$\lim_{N\to\infty}\frac{E}{N} = \varepsilon^{GP}, \tag{10}$$

and for the reduced density matrices [67, 68] per particle

$$\lim_{N\to\infty}\frac{\rho_1(\mathbf{x},\mathbf{x}')}{N_1} = \rho_1^{GP}(\mathbf{x},\mathbf{x}'), \qquad \lim_{N\to\infty}\frac{\rho_2(\mathbf{y},\mathbf{y}')}{N_2} = \rho_2^{GP}(\mathbf{y},\mathbf{y}'), \tag{11}$$

and

$$\lim_{N\to\infty}\frac{\rho_{12}(\mathbf{x},\mathbf{x}',\mathbf{y},\mathbf{y}')}{N_1N_2} = \rho_1^{GP}(\mathbf{x},\mathbf{x}')\rho_2^{GP}(\mathbf{y},\mathbf{y}'). \tag{12}$$

The reduced one-particle density matrices at the Gross-Pitaevskii level are given by $\rho_1^{GP}(\mathbf{x},\mathbf{x}') = \phi_1^{GP}(\mathbf{x})\left\{\phi_1^{GP}(\mathbf{x}')\right\}^*$ and $\rho_2^{GP}(\mathbf{y},\mathbf{y}') = \phi_2^{GP}(\mathbf{y})\left\{\phi_2^{GP}(\mathbf{y}')\right\}^*$. The question we would like to address in this work is in what capacity (i.e., for which quantities and by how much) the many-body and mean-field solutions are different, even when each of the species in the mixture is 100% condensed [Eq. (11)].

To proceed, we need to express the Gross-Pitaevskii wavefunction (7) via the Jacobi coordinates (2). Using (A.1) and (A.2) we have

$$\begin{aligned}
\Phi^{GP}(\mathbf{Q}_1,\ldots,\mathbf{Q}_N) &= \left(\frac{m_1\Omega_1^{GP}}{\pi}\right)^{\frac{3N_1}{4}}\left(\frac{m_2\Omega_2^{GP}}{\pi}\right)^{\frac{3N_1}{4}} \times \\
&\times e^{-\frac{m_1}{2}\Omega_1^{GP}\left[\sum_{s=1}^{N_1-1}\mathbf{Q}_s^2+\left(\frac{\sqrt{N_2}m_2}{M}\mathbf{Q}_{N-1}+\sqrt{N_1}\mathbf{Q}_N\right)^2\right]} e^{-\frac{m_2}{2}\Omega_2^{GP}\left[\sum_{s=N_1}^{N-2}\mathbf{Q}_s^2+\left(-\frac{\sqrt{N_1}m_1}{M}\mathbf{Q}_{N-1}+\sqrt{N_2}\mathbf{Q}_N\right)^2\right]} = \\
&= \left(\frac{m_1\Omega_1^{GP}}{\pi}\right)^{\frac{3N_1}{4}}\left(\frac{m_2\Omega_2^{GP}}{\pi}\right)^{\frac{3N_1}{4}} e^{-\frac{m_1}{2}\Omega_1^{GP}\sum_{s=1}^{N_1-1}\mathbf{Q}_s^2}\, e^{-\frac{m_2}{2}\Omega_2^{GP}\sum_{s=N_1}^{N-2}\mathbf{Q}_s^2} \times \\
&\times e^{-\frac{M_{12}}{2}\left[\frac{m_2N_2}{M}\Omega_1^{GP}+\frac{m_1N_1}{M}\Omega_2^{GP}\right]\mathbf{Q}_{N-1}^2} e^{-\frac{M}{2}\left[\frac{m_1N_1}{M}\Omega_1^{GP}+\frac{m_2N_2}{M}\Omega_2^{GP}\right]\mathbf{Q}_N^2} e^{-M_{12}\sqrt{N_1N_2}\left[\Omega_1^{GP}-\Omega_2^{GP}\right]\mathbf{Q}_{N-1}\mathbf{Q}_N}.
\end{aligned} \tag{13}$$

We see that the Gross-Pitaevskii wavefunction is 'nearly' separable in terms of the Jacobi coordinates, except of the last term which couples $\mathbf{Q}_{N-1}$ and $\mathbf{Q}_N$. Interestingly, along the sector in parameter space in which the mean-field frequencies [see (7)] are equal,

$$\sqrt{\omega^2+\frac{2}{m_1}(\Lambda_1+\Lambda_{21})} = \sqrt{\omega^2+\frac{2}{m_2}(\Lambda_2+\Lambda_{12})} \quad\Longleftrightarrow\quad m_2(\Lambda_1+\Lambda_{21}) = m_1(\Lambda_2+\Lambda_{12}), \tag{14}$$

the Gross-Pitaevskii wavefunction is separable in terms of the Jacobi coordinates, as does the exact wavefunction. Relation (14) defines an interaction–mass balance condition between the two species for the separability of the center-of-mass coordinate of the mixture at the mean-field level. Surprisingly, this condition for separability is required even in the case the inter-species interaction is zero. In other words, the center-of-mass coordinate $\mathbf{Q}_N$ of two (harmonic-interacting) trapped BECs is not separable at the Gross-Pitaevskii level, even if the two species do not interact with each other, unless $m_2\Lambda_1 = m_1\Lambda_2$. We note that condition (14) obviously holds for the symmetric mixture [42]. On the other hand, the Gross-Pitaevskii wavefunction is separable in terms of the center-of-mass coordinates of the individual species (see below), whereas the exact wavefunction is not.

We now express the position and momentum center-of-mass operators of the mixture and of each of the species in terms of the Jacobi coordinates. Thus we have

$$\hat{\mathbf{R}}_{CM} = \mathbf{Q}_N, \quad \hat{\mathbf{P}}_{CM} = \frac{1}{i}\frac{\partial}{\partial \mathbf{Q}_N}, \quad [\hat{\mathbf{R}}_{CM}, \hat{\mathbf{P}}_{CM}] = \boldsymbol{i} \quad \forall N \tag{15}$$

for the mixture ($\boldsymbol{i}$ and similarly $\mathbf{1}$ below are shorthand symbols for i and 1 in each of the three Cartesian components) and

$$\begin{aligned}
&\hat{\mathbf{X}}_{CM} = \sqrt{\frac{N_2}{N_1}\frac{m_2}{M}}\mathbf{Q}_{N-1} + \mathbf{Q}_N, \quad \hat{\mathbf{P}}_{X_{CM}} = \frac{1}{i}\frac{\partial}{\partial \mathbf{X}_{CM}} = \sqrt{N_1 N_2}\frac{1}{i}\frac{\partial}{\partial \mathbf{Q}_{N-1}} + \frac{N_1 m_1}{M}\frac{1}{i}\frac{\partial}{\partial \mathbf{Q}_N}, \\
&[\hat{\mathbf{X}}_{CM}, \hat{\mathbf{P}}_{\mathbf{X}_{CM}}] = \boldsymbol{i}, \quad \forall N, \\
&\hat{\mathbf{Y}}_{CM} = -\sqrt{\frac{N_1}{N_2}\frac{m_1}{M}}\mathbf{Q}_{N-1} + \mathbf{Q}_N, \quad \hat{\mathbf{P}}_{Y_{CM}} = \frac{1}{i}\frac{\partial}{\partial \mathbf{Y}_{CM}} = -\sqrt{N_1 N_2}\frac{1}{i}\frac{\partial}{\partial \mathbf{Q}_{N-1}} + \frac{N_2 m_2}{M}\frac{1}{i}\frac{\partial}{\partial \mathbf{Q}_N}, \\
&[\hat{\mathbf{Y}}_{CM}, \hat{\mathbf{P}}_{\mathbf{Y}_{CM}}] = \boldsymbol{i}, \quad \forall N,
\end{aligned} \tag{16}$$

for each of the species. The center-of-mass operators of the species and mixture are related by the relations

$$\frac{m_1 N_1 \hat{\mathbf{X}}_{CM} + m_2 N_2 \hat{\mathbf{Y}}_{CM}}{M} = \hat{\mathbf{R}}_{CM}, \quad \hat{\mathbf{P}}_{X_{CM}} + \hat{\mathbf{P}}_{Y_{CM}} = \hat{\mathbf{P}}_{CM} \tag{17}$$

which are to be used in the computation of the variances at the mean-field level.

In what follows we use a compact notation to denote the variance (of the three Cartesian components) of an operator $\hat{\mathbf{O}} = (\hat{O}_1, \hat{O}_2, \hat{O}_3)$ with respect to the exact and Gross-Pitaevskii wavefunctions,

$$\Delta^2_{\hat{\mathbf{O}}} = \langle\Psi|\hat{\mathbf{O}}^2|\Psi\rangle - \langle\Psi|\hat{\mathbf{O}}|\Psi\rangle^2, \qquad \Delta^2_{\hat{\mathbf{O}},GP} = \langle\Phi^{GP}|\hat{\mathbf{O}}^2|\Phi^{GP}\rangle - \langle\Phi^{GP}|\hat{\mathbf{O}}|\Phi^{GP}\rangle^2, \tag{18}$$

where $\hat{\mathbf{O}}^2 \equiv (\hat{O}_1^2, \hat{O}_2^2, \hat{O}_3^2)$.

The variances of the center-of-mass position and momentum operators of the mixture, at the exact many-body level, are given by

$$\Delta^2_{\hat{\mathbf{R}}_{CM}} = \frac{1}{M}\frac{1}{2\omega}\mathbf{1}, \qquad \Delta^2_{\hat{\mathbf{P}}_{CM}} = M\frac{\omega}{2}\mathbf{1}, \quad \forall N. \tag{19}$$

The position variance decreases with the mass of the mixture, whereas the momentum variance increases with the mass. Otherwise, they do not depend on other parameters of the mixture, like the interaction parameters for instance. Consequently, their uncertainty product

$$\Delta^2_{\hat{\mathbf{R}}_{CM}}\Delta^2_{\hat{\mathbf{P}}_{CM}} = \frac{1}{4}\mathbf{1}, \quad \forall N \tag{20}$$

holds for any number of particles. The uncertainty product is minimal, reflecting the separability of the center-of-mass of the mixture in the harmonic trap.

On the other hand, at the Gross-Pitaevskii level we find

$$\Delta^2_{\hat{\mathbf{R}}_{CM},GP} = \frac{1}{M}\left[\frac{m_1N_1}{M}\frac{1}{\sqrt{1+\frac{2}{m_1\omega^2}(\Lambda_1+\Lambda_{21})}} + \frac{m_2N_2}{M}\frac{1}{\sqrt{1+\frac{2}{m_2\omega^2}(\Lambda_2+\Lambda_{12})}}\right]\frac{1}{2\omega}\mathbf{1}, \tag{21}$$

$$\Delta^2_{\hat{\mathbf{P}}_{\mathbf{R}_{CM}},GP} = M\left[\frac{m_1N_1}{M}\sqrt{1+\frac{2}{m_1\omega^2}(\Lambda_1+\Lambda_{21})} + \frac{m_2N_2}{M}\sqrt{1+\frac{2}{m_2\omega^2}(\Lambda_2+\Lambda_{12})}\right]\frac{\omega}{2}\mathbf{1}, \quad \forall N.$$

The center-of-mass momentum (position) variance again scales with the (inverse) mass of the system, but depends explicitly on the products of the relative mass of each species times the corresponding (inverse) Gross-Pitaevskii frequency. The variances depend explicitly on all parameters of the mixture, in particular on the interaction parameters. This is entirely different than the many-body dependence (19), and reflects the dressing of the frequency (and inseparability) of the center-of-mass coordinate at the mean-field level. Consequently, the uncertainty product

$$\Delta^2_{\hat{\mathbf{R}}_{CM},GP}\Delta^2_{\hat{\mathbf{P}}_{CM},GP} = \left\{1 + \frac{m_1m_2\Lambda_{12}\Lambda_{21}}{(m_1\Lambda_{12}+m_2\Lambda_{21})^2}\left[\frac{\sqrt{1+\frac{2}{m_1\omega^2}(\Lambda_1+\Lambda_{21})}}{\sqrt{1+\frac{2}{m_2\omega^2}(\Lambda_2+\Lambda_{12})}} + \right.\right.$$
$$\left.\left. + \frac{\sqrt{1+\frac{2}{m_2\omega^2}(\Lambda_2+\Lambda_{12})}}{\sqrt{1+\frac{2}{m_1\omega^2}(\Lambda_1+\Lambda_{21})}} - 2\right]\right\}\frac{1}{4}\mathbf{1} \geq \frac{1}{4}\mathbf{1}, \quad \forall N \tag{22}$$

is larger than the minimal, except when the interaction–mass balance condition (14) holds and the Gross-Pitaevskii wavefunction (13) becomes separable with respect to the center-of-mass coordinate. For instance, increasing either of the intra-species interaction parameters Λ_1 or Λ_2 leads to an increase of the uncertainty product. The above described discrepancy between the mean-field and exact results can be used as a characterization and measure of the many-body contribution to the physics of a given 100%-condensed trapped mixture.

We now move to the properties of the center-of-mass operators of the species. Using relations (16) in terms of the Jacobi coordinates we find

$$\Delta^2_{\hat{\mathbf{X}}_{CM}} = \frac{1}{m_1N_1}\frac{1}{2\omega}\left[\frac{m_1N_1}{M} + \frac{m_2N_2}{M}\frac{1}{\sqrt{1+\frac{2}{\omega^2}\left(\frac{\Lambda_{12}}{m_2}+\frac{\Lambda_{21}}{m_1}\right)}}\right]\mathbf{1},$$

$$\Delta^2_{\hat{\mathbf{P}}_{\mathbf{X}_{CM}}} = m_1N_1\frac{\omega}{2}\left[\frac{m_1N_1}{M} + \frac{m_2N_2}{M}\sqrt{1+\frac{2}{\omega^2}\left(\frac{\Lambda_{12}}{m_2}+\frac{\Lambda_{21}}{m_1}\right)}\right]\mathbf{1}, \quad \forall N,$$

$$\Delta^2_{\hat{\mathbf{Y}}_{CM}} = \frac{1}{m_2N_2}\frac{1}{2\omega}\left[\frac{m_2N_2}{M} + \frac{m_1N_1}{M}\frac{1}{\sqrt{1+\frac{2}{\omega^2}\left(\frac{\Lambda_{12}}{m_2}+\frac{\Lambda_{21}}{m_1}\right)}}\right]\mathbf{1},$$

$$\Delta^2_{\hat{\mathbf{P}}_{\mathbf{Y}_{CM}}} = m_2N_2\frac{\omega}{2}\left[\frac{m_2N_2}{M} + \frac{m_1N_1}{M}\sqrt{1+\frac{2}{\omega^2}\left(\frac{\Lambda_{12}}{m_2}+\frac{\Lambda_{21}}{m_1}\right)}\right]\mathbf{1}, \quad \forall N. \tag{23}$$

The center-of-mass position and momentum variances of each species depend only on the inter-species interaction parameters Λ_{12} and Λ_{21}, and not on the intra-species parameters Λ_1 and

Λ_2. Increasing the inter-species interaction, such that the species attract each other more, decreases (increases) the position (momentum) variance, and vice versa. Generally, the variances of species 1 are different than those of species 2 because of the mass–particle imbalance (i.e., $m_1N_1 \neq m_2N_2$) of the species. Interestingly, however, the two uncertainty products are equal

$$\Delta^2_{\hat{\mathbf{X}}_{CM}}\Delta^2_{\hat{\mathbf{P}}_{\mathbf{X}_{CM}}} = \Delta^2_{\hat{\mathbf{Y}}_{CM}}\Delta^2_{\hat{\mathbf{P}}_{\mathbf{Y}_{CM}}} = \left\{1 + \frac{m_1m_2\Lambda_{12}\Lambda_{21}}{(m_1\Lambda_{12}+m_2\Lambda_{21})^2}\left[\frac{1}{\sqrt{1+\frac{2}{\omega^2}\left(\frac{\Lambda_{12}}{m_2}+\frac{\Lambda_{21}}{m_1}\right)}} + \right.\right.$$
$$\left.\left. + \sqrt{1+\frac{2}{\omega^2}\left(\frac{\Lambda_{12}}{m_2}+\frac{\Lambda_{21}}{m_1}\right)} - 2\right]\right\}\frac{1}{4}\mathbf{1} > \frac{1}{4}\mathbf{1}, \quad \forall N. \tag{24}$$

The uncertainty products of the individual species are always larger than (the minimal uncertainty product) $\frac{1}{4}$, unless the inter-species interaction is zero, and we have two decoupled systems. The dependence on the mass–particle imbalance is then absorbed. In the specific case of a symmetric mixture, these expressions boil down to those given in [42].

At the mean-field level, the Gross-Pitaevskii wavefunction (7,13) is separable in terms of the center-of-mass coordinates of each species and we thus have

$$\Delta^2_{\hat{\mathbf{X}}_{CM},GP} = \frac{1}{m_1N_1}\frac{1}{2\omega}\frac{1}{\sqrt{1+\frac{2}{m_1\omega^2}(\Lambda_1+\Lambda_{21})}}\mathbf{1},$$
$$\Delta^2_{\hat{\mathbf{P}}_{\mathbf{X}_{CM}},GP} = m_1N_1\frac{\omega}{2}\sqrt{1+\frac{2}{m_1\omega^2}(\Lambda_1+\Lambda_{21})}\mathbf{1}, \quad \forall N,$$
$$\Delta^2_{\hat{\mathbf{Y}}_{CM},GP} = \frac{1}{m_2N_2}\frac{1}{2\omega}\frac{1}{\sqrt{1+\frac{2}{m_2\omega^2}(\Lambda_2+\Lambda_{12})}}\mathbf{1},$$
$$\Delta^2_{\hat{\mathbf{P}}_{\mathbf{Y}_{CM}},GP} = m_2N_2\frac{\omega}{2}\sqrt{1+\frac{2}{m_2\omega^2}(\Lambda_2+\Lambda_{12})}\mathbf{1}, \quad \forall N. \tag{25}$$

Again, the center-of-mass position and momentum variances of species 1 are generally different than those of species 2. However, at the mean-field level they depend on both the intra-species and inter-species interaction parameters, rather than only on the inter-species interaction parameters. Furthermore, the (non-trivial) dependence on the mass–particle imbalance is absent. The two uncertainty products are again equal,

$$\Delta^2_{\hat{\mathbf{X}}_{CM},GP}\Delta^2_{\hat{\mathbf{P}}_{\mathbf{X}_{CM}},GP} = \Delta^2_{\hat{\mathbf{Y}}_{CM},GP}\Delta^2_{\hat{\mathbf{P}}_{\mathbf{Y}_{CM}},GP} = \frac{1}{4}\mathbf{1}, \quad \forall N, \tag{26}$$

but, in contrast to the exact relation (24), are actually minimal. This is because of the (artificial) separation of the center-of-mass coordinates of the individual species in the Gross-Pitaevskii wavefunction. This result adds to our above conclusion, i.e., that in addition to the uncertainty product of the whole mixture also the uncertainty products of the individual species can be used for characterizing and defining the many-body contributions to the physics of a 100%-condensed trapped mixture.

We now move to discuss the variance of the angular momentum in the mixture at both the exact and mean-field levels. To identify the contribution to the angular momentum at the level of a single particle, individual species, and the whole mixture,

$$\hat{\mathbf{l}}_{1,i'} = \frac{1}{i}\mathbf{x}_{i'}\times\frac{\partial}{\partial\mathbf{x}_{i'}}, \quad \hat{\mathbf{l}}_{2,j} = \frac{1}{i}\mathbf{y}_j\times\frac{\partial}{\partial\mathbf{y}_j}, \quad \hat{\mathbf{L}}_1 = \sum_{i=1}^{N_1}\hat{\mathbf{l}}_{1,i}, \quad \hat{\mathbf{L}}_2 = \sum_{j=1}^{N_2}\hat{\mathbf{l}}_{2,j}, \quad \hat{\mathbf{L}} = \hat{\mathbf{L}}_1+\hat{\mathbf{L}}_2, \tag{27}$$

it is instrumental to exploit the representation of the wavefunctions in the laboratory frame (Cartesian coordinates). The mean-field wavefunction (7) is a 'spherically symmetric' double-product state. It hence satisfies at the level of a single particle $\hat{\mathbf{l}}_{1,i}\Phi^{GP} = \mathbf{0}\Phi^{GP}$ and $\hat{\mathbf{l}}_{2,j}\Phi^{GP} = \mathbf{0}\Phi^{GP}$. Consequently,

$$\Delta^2_{\hat{\mathbf{L}}_1,GP} = \Delta^2_{\hat{\mathbf{L}}_2,GP} = \Delta^2_{\hat{\mathbf{L}},GP} = \mathbf{0}, \quad \forall N. \tag{28}$$

For the exact wavefunction we recall its structure in the laboratory frame [43]

$$\Psi(\mathbf{x}_1,\ldots,\mathbf{x}_{N_1},\mathbf{y}_1,\ldots,\mathbf{y}_{N_2}) = \left(\frac{m_1\Omega_1}{\pi}\right)^{\frac{3(N_1-1)}{4}} \left(\frac{m_2\Omega_2}{\pi}\right)^{\frac{3(N_2-1)}{4}} \left(\frac{M_{12}\Omega_{12}}{\pi}\right)^{\frac{3}{4}} \left(\frac{M\omega}{\pi}\right)^{\frac{3}{4}} \times$$
$$\times e^{-\frac{\alpha_1}{2}\sum_{j=1}^{N_1}\mathbf{x}_j^2-\beta_1\sum_{1\le j<k}^{N_1}\mathbf{x}_j\cdot\mathbf{x}_k} e^{-\frac{\alpha_2}{2}\sum_{j=1}^{N_2}\mathbf{y}_j^2-\beta_2\sum_{1\le j<k}^{N_2}\mathbf{y}_j\cdot\mathbf{y}_k} e^{+\gamma\sum_{j=1}^{N_1}\sum_{k=1}^{N_2}\mathbf{x}_j\cdot\mathbf{y}_k}, \tag{29}$$

where the inter-species coupling constant is

$$\gamma = M_{12}(\Omega_{12}-\omega) = \frac{\omega}{\frac{N_1}{m_2}+\frac{N_2}{m_1}}\left(\sqrt{1+\frac{2}{\omega^2}\left(\frac{\Lambda_{12}}{m_2}+\frac{\Lambda_{21}}{m_1}\right)}-1\right). \tag{30}$$

The other constants α_1, β_1, α_2, and β_2 appearing in (29) are not needed for the computation of the angular momentum and given in [43]. Examining the structure of (29) we see that the α_1, α_2 terms alone are eigenfunctions of $\hat{\mathbf{l}}_{1,i}$ and $\hat{\mathbf{l}}_{2,j}$, respectively, the β_1, β_2 terms alone are eigenfunctions of $\hat{\mathbf{L}}_1$ and $\hat{\mathbf{L}}_2$, respectively, and the γ term is an eigenfunction of $\hat{\mathbf{L}}$. All in all, whereas $\hat{\mathbf{L}}\Psi = \mathbf{0}\Psi$, only $\langle\Psi|\hat{\mathbf{l}}_{1,i}|\Psi\rangle = 0$, $\langle\Psi|\hat{\mathbf{l}}_{2,j}|\Psi\rangle = 0$ and $\langle\Psi|\hat{\mathbf{L}}_1|\Psi\rangle = 0$ and $\langle\Psi|\hat{\mathbf{L}}_2|\Psi\rangle = 0$ hold for the exact ground state. This is unlike the structure and above properties of the Gross-Pitaevskii wavefunction (7). Of course,

$$\Delta^2_{\hat{\mathbf{L}}} = \mathbf{0}, \quad \forall N. \tag{31}$$

Since Ψ is not an eigenfunction of either $\hat{\mathbf{L}}_1$ or $\hat{\mathbf{L}}_2$ (unless the inter-species interaction is zero and the species decoupled), fluctuations are expected. Making use of the structure of (29) we have

$$\hat{\mathbf{L}}_1\Psi = -\hat{\mathbf{L}}_2\Psi = \frac{1}{i}\gamma N_1N_2(\hat{\mathbf{X}}_{CM}\times\hat{\mathbf{Y}}_{CM})\Psi =$$
$$= \frac{1}{i}\gamma N_1N_2\left[\left(\sqrt{\frac{N_2}{N_1}\frac{m_2}{M}}\mathbf{Q}_{N-1}+\mathbf{Q}_N\right)\times\left(-\sqrt{\frac{N_1}{N_2}\frac{m_1}{M}}\mathbf{Q}_{N-1}+\mathbf{Q}_N\right)\right]\Psi =$$
$$= \frac{1}{i}\gamma\sqrt{N_1N_2}(\mathbf{Q}_{N-1}\times\mathbf{Q}_N)\Psi. \tag{32}$$

Thus, for the variance of the intra-species angular momentum operators $\hat{\mathbf{L}}_1$ and $\hat{\mathbf{L}}_2$ we find

$$\Delta^2_{\hat{\mathbf{L}}_1} = \Delta^2_{\hat{\mathbf{L}}_2} = \gamma^2N_1N_2\langle\Psi|(\mathbf{Q}_{N-1}\times\mathbf{Q}_N)^2|\Psi\rangle =$$
$$= \frac{1}{2}\frac{\frac{\Lambda_{12}}{m_2}\frac{\Lambda_{21}}{m_1}}{\left(\frac{\Lambda_{12}}{m_2}+\frac{\Lambda_{21}}{m_1}\right)^2}\frac{\left[\sqrt{1+\frac{2}{\omega^2}\left(\frac{\Lambda_{12}}{m_2}+\frac{\Lambda_{21}}{m_1}\right)}-1\right]^2}{\sqrt{1+\frac{2}{\omega^2}\left(\frac{\Lambda_{12}}{m_2}+\frac{\Lambda_{21}}{m_1}\right)}}\mathbf{1}, \quad \forall N, \tag{33}$$

where we have used the separability of Ψ in terms of the Jacobi coordinates and the spherical symmetry of the problem. The variance is independent of the number of particles and is always non-zero for non-zero inter-species interaction. On the other hand, if we examine the variances of the intra-species angular-momentum operators per particle, $\frac{\hat{\mathbf{L}}_1}{N_1}$ and $\frac{\hat{\mathbf{L}}_2}{N_2}$, we find from $\Delta^2_{\frac{\hat{\mathbf{L}}_1}{N_1}} = \frac{1}{N_1^2}\Delta^2_{\hat{\mathbf{L}}_1}$, $\Delta^2_{\frac{\hat{\mathbf{L}}_2}{N_2}} = \frac{1}{N_2^2}\Delta^2_{\hat{\mathbf{L}}_2}$ that they vanish in the infinite-particle limit. In other words, at the single-particle level the angular-momentum each boson carries in the infinite-particle limit vanishes, just like the mean-field behavior. It would be instructive to study then the variance of the appropriate conjugate angular-momentum (angle) variable [69] in the mixture, which is left for further investigations.

The exact and Gross-Pitaevskii wavefunctions are given analytically above for any number N_1, N_2 of particles in the mixture, and are obviously different from each other even in the limit N goes to infinity. Motivated by recent results in the single-species case [12, 14], we would like to study the overlap between the Gross-Pitaevskii and exact wavefunctions which, by virtue of their difference, must be smaller than 1.

For the final result of the overlap we find

$$S_{12}(N_1,N_2,\Lambda_1,\Lambda_2,\Lambda_{12},\Lambda_{21}) = \langle\Phi^{GP}|\Psi\rangle =$$
$$= 2^{\frac{3}{2}N}\frac{\Omega_1^{\frac{3}{4}(N_1-1)}[\Omega_1^{GP}]^{\frac{3}{4}(N_1-1)}}{\left[\Omega_1+\Omega_1^{GP}\right]^{\frac{3}{2}(N_1-1)}}\frac{\Omega_2^{\frac{3}{4}(N_2-1)}[\Omega_2^{GP}]^{\frac{3}{4}(N_2-1)}}{\left[\Omega_2+\Omega_2^{GP}\right]^{\frac{3}{2}(N_2-1)}}\times \tag{34}$$
$$\times\frac{[\Omega_1^{GP}]^{\frac{3}{4}}[\Omega_2^{GP}]^{\frac{3}{4}}\Omega_{12}^{\frac{3}{4}}\omega^{\frac{3}{4}}}{\left[\left(\Omega_{12}+\frac{m_2N_2}{M}\Omega_1^{GP}+\frac{m_1N_1}{M}\Omega_2^{GP}\right)\left(\omega+\frac{m_1N_1}{M}\Omega_1^{GP}+\frac{m_2N_2}{M}\Omega_2^{GP}\right)-\frac{m_1N_1m_2N_2}{(m_1N_1+m_2N_2)^2}\left(\Omega_1^{GP}-\Omega_2^{GP}\right)^2\right]^{\frac{3}{2}}}.$$

The overlap depends on all interaction parameters in the mixture and explicitly on the number of particles in each species. We can now perform the infinite-particle limit. The final result reads

$$\lim_{N\to\infty} S_{12}(N_1,N_2,\Lambda_1,\Lambda_2,\Lambda_{12},\Lambda_{21}) =$$
$$= 2^3\frac{\left[1+\frac{2}{m_1\omega^2}(\Lambda_1+\Lambda_{21})\right]^{\frac{3}{8}}\left[1+\frac{2}{m_2\omega^2}(\Lambda_2+\Lambda_{12})\right]^{\frac{3}{8}}\left[1+\frac{2}{\omega^2}\left(\frac{\Lambda_{12}}{m_2}+\frac{\Lambda_{21}}{m_1}\right)\right]^{\frac{3}{8}}}{[AB-C^2]^{\frac{3}{2}}} \tag{35}$$

with

$$A = \left[1+\frac{2}{\omega^2}\left(\frac{\Lambda_{12}}{m_2}+\frac{\Lambda_{21}}{m_1}\right)\right]+\frac{m_2\Lambda_{21}}{m_1\Lambda_{12}+m_2\Lambda_{21}}\left[1+\frac{2}{m_1\omega^2}(\Lambda_1+\Lambda_{21})\right]+$$
$$+\frac{m_1\Lambda_{12}}{m_1\Lambda_{12}+m_2\Lambda_{21}}\left[1+\frac{2}{m_2\omega^2}(\Lambda_2+\Lambda_{12})\right],$$
$$B = 1+\frac{m_1\Lambda_{12}}{m_1\Lambda_{12}+m_2\Lambda_{21}}\left[1+\frac{2}{m_1\omega^2}(\Lambda_1+\Lambda_{21})\right]+\frac{m_2\Lambda_{21}}{m_1\Lambda_{12}+m_2\Lambda_{21}}\left[1+\frac{2}{m_2\omega^2}(\Lambda_2+\Lambda_{12})\right],$$
$$C = \frac{\sqrt{m_1m_2\Lambda_{12}\Lambda_{21}}}{m_1\Lambda_{12}+m_2\Lambda_{21}}\left\{\left[1+\frac{2}{m_1\omega^2}(\Lambda_1+\Lambda_{21})\right]-\left[1+\frac{2}{m_2\omega^2}(\Lambda_2+\Lambda_{12})\right]\right\}. \tag{36}$$

At the infinite-particle limit, the overlap depends on the interaction parameters only and, as expected, is always smaller than 1 in presence of either intra-species or inter-species interactions. At this limit, the overlap boils down to a product of two single-species overlaps (at the infinite-particle limit) provided there is no inter-species interaction *and* then only when the interaction–mass balance condition (14) holds and the mean-field frequencies are equal. This separability relation between the overlap in the mixture and the overlaps within each of its species, in the infinite-particle limit, reads $S_{12}(\Lambda_1,\frac{m_2}{m_1}\Lambda_1,0,0) = S_1(\Lambda_1)S_2(\frac{m_2}{m_1}\Lambda_1)$, with obvious notation for the quantities. This concludes our study.

4. Summary

We have considered in the present work a generic trapped mixture of Bose-Einstein condensates whose ground-state wavefunction can be prescribed analytically at the exact and mean-field levels. This situation greatly facilitates a transparent and comparative study of properties of the mixture at both levels of theory. We have derived in this work general expressions for any number of particles in the mixture, and have concentrated on the infinite-particle limit in which both wavefunctions admit the same energy per particle and 100% condensation of each of the species.

We have computed and investigated the center-of-mass position and momentum variances, their uncertainty product, and the angular-momentum variance of each of the species in the mixture as well as of the whole mixture. Particular attention has been paid at identifying how differences between the exact and mean-field wavefunctions lead to explicit deviations between the respective quantities. The renormalization of the center-of-mass frequency and that of the relative coordinate between the center-of-mass of each species, as well as the (artificial) coupling between these two Jacobi coordinates and the (artificial) decoupling between the center-of-mass of each species, all in the Gross-Pitaevskii wavefunction, lead to discrepancies between the exact and mean-field properties. At the bottom line, the two different wavefunctions must have a lower than 1 overlap, which has been computed explicitly and analytically as well. The results obtained in this study can be considered as a step forward in characterizing how important are many-body effects in a generic trapped 100%-condensed bosonic mixture.

As a brief outlook, it would be interesting to study more involved uncertainty relations [70, 71] in a generic mixture, to extend the investigations on the relation of the exact and mean-field solutions to excited states, when possible, and to explore the further opportunities hiding in trapped multi-species mixtures.

Acknowledgments

This research was supported by the Israel Science Foundation (Grant No. 600/15). We thank Kaspar Sakmann and Lorenz Cederbaum for discussions. OEA is grateful to the continuous hospitality of the Lewiner Institute for Theoretical Physics (LITP) at the Department of Physics, Technion.

Appendix A. Transformation between the quadratic terms in Cartesian coordinates (laboratory frame) and Jacobi coordinates

The position terms are related by

$$\sum_{i=1}^{N_1} \mathbf{x}_i^2 = \sum_{s=1}^{N_1-1} \mathbf{Q}_s^2 + \left(\frac{\sqrt{N_2}m_2}{M}\mathbf{Q}_{N-1} + \sqrt{N_1}\mathbf{Q}_N\right)^2 =$$
$$= \sum_{s=1}^{N_1-1} \mathbf{Q}_s^2 + \frac{N_2 m_2^2}{M^2}\mathbf{Q}_{N-1}^2 + N_1\mathbf{Q}_N^2 + \frac{2\sqrt{N_1 N_2}m_2}{M}\mathbf{Q}_{N-1}\mathbf{Q}_N,$$
$$\sum_{j=1}^{N_2} \mathbf{y}_j^2 = \sum_{s=N_1}^{N-2} \mathbf{Q}_s^2 + \left(-\frac{\sqrt{N_1}m_1}{M}\mathbf{Q}_{N-1} + \sqrt{N_2}\mathbf{Q}_N\right)^2 =$$
$$= \sum_{s=N_1}^{N-2} \mathbf{Q}_s^2 + \frac{N_1 m_1^2}{M^2}\mathbf{Q}_{N-1}^2 + N_2\mathbf{Q}_N^2 - \frac{2\sqrt{N_1 N_2}m_1}{M}\mathbf{Q}_{N-1}\mathbf{Q}_N \tag{A.1}$$

and the momentum terms by

$$\sum_{i=1}^{N_1} \frac{\partial^2}{\partial \mathbf{x}_i^2} = \sum_{s=1}^{N_1-1} \frac{\partial^2}{\partial \mathbf{Q}_s^2} + \left(\sqrt{N_2}\frac{\partial}{\partial \mathbf{Q}_{N-1}} + \frac{\sqrt{N_1}m_1}{M}\frac{\partial}{\partial \mathbf{Q}_N}\right)^2 =$$
$$= \sum_{s=1}^{N_1-1} \frac{\partial^2}{\partial \mathbf{Q}_s^2} + N_2\frac{\partial^2}{\partial \mathbf{Q}_{N-1}^2} + \frac{N_1 m_1^2}{M^2}\frac{\partial^2}{\partial \mathbf{Q}_N^2} + \frac{2\sqrt{N_1N_2}m_1}{M}\frac{\partial}{\partial \mathbf{Q}_{N-1}}\frac{\partial}{\partial \mathbf{Q}_N},$$
$$\sum_{j=1}^{N_2} \frac{\partial^2}{\partial \mathbf{y}_j^2} = \sum_{s=N_1}^{N-2} \frac{\partial^2}{\partial \mathbf{Q}_s^2} + \left(-\sqrt{N_1}\frac{\partial}{\partial \mathbf{Q}_{N-1}} + \frac{\sqrt{N_2}m_2}{M}\frac{\partial}{\partial \mathbf{Q}_N}\right)^2 =$$
$$= \sum_{s=N_1}^{N-2} \frac{\partial^2}{\partial \mathbf{Q}_s^2} + N_1\frac{\partial^2}{\partial \mathbf{Q}_{N-1}^2} + \frac{N_2 m_2^2}{M^2}\frac{\partial^2}{\partial \mathbf{Q}_N^2} - \frac{2\sqrt{N_1N_2}m_2}{M}\frac{\partial}{\partial \mathbf{Q}_{N-1}}\frac{\partial}{\partial \mathbf{Q}_N}. \quad \text{(A.2)}$$

References

[1] Anderson M H, Ensher J R, Matthews M R, Wieman C E and Cornell E A 1995 *Science* **269** 198
[2] Bradley C C, Sackett C A, Tollett J J and Hulet R G 1995 *Phys. Rev. Lett.* **75** 1687
[3] Davis K B, Mewes M O, Andrews M R, van Druten N J, Durfee D S, Kurn D M and Ketterle W 1995 *Phys. Rev. Lett.* **75** 3969
[4] Castin Y and Dum R 1998 *Phys. Rev. A* **57** 3008
[5] Lieb E H, Seiringer R and Yngvason J 2000 *Phys. Rev. A* **61** 043602
[6] Lieb E H and Seiringer R 2002 *Phys. Rev. Lett.* **88** 170409
[7] Erdős L, Schlein B and Yau H T 2007 *Phys. Rev. Lett.* **98** 040404
[8] Erdős L, Schlein B and Yau H T 2007 *Invent. Math.* **167** 515
[9] Yukalov V I 2011 *Phys. Part. Nucl.* **42** 460
[10] Klaiman S and Alon O E 2015 *Phys. Rev. A* **91** 063613
[11] Klaiman S, Streltsov A I and Alon O E 2016 *Phys. Rev. A* **93** 023605
[12] Klaiman S and Cederbaum L S 2016 *Phys. Rev. A* **94** 063648
[13] Klaiman S, Streltsov A I and Alon O E 2017 *J. Phys.: Conf. Ser.* **826** 012020
[14] Cederbaum L S 2017 *Phys. Rev. A* **96** 013615
[15] Myatt C J, Burt E A, Ghrist R W, Cornell E A and Wieman C E 1997 *Phys. Rev. Lett.* **78** 586
[16] Stamper-Kurn D M, Andrews M R, Chikkatur A P, Inouye S, Miesner H-J, Stenger J and Ketterle W 1998 *Phys. Rev. Lett.* **80** 2027
[17] Ho T-L and Shenoy V B 1996 *Phys. Rev. Lett.* **77** 3276
[18] Esry B D, Greene C H, Burke J P Jr and Bohn J L 1997 *Phys. Rev. Lett.* **78** 3594
[19] Pu H and Bigelow N P 1998 *Phys. Rev. Lett.* **80** 1130
[20] Timmermans E 1998 *Phys. Rev. Lett.* **81** 5718
[21] Eckardt A, Weiss C and Holthaus M 2004 *Phys. Rev. A* 70 043615
[22] Alon O E, Streltsov A I and Cederbaum L S 2007 *Phys. Rev. A* **76** 062501
[23] Sakhel A R, DuBois J L and Glyde H R 2008 *Phys. Rev. A* **77** 043627
[24] Oleś B and Sacha K 2008 *J. Phys. A* **41** 145005
[25] Hao Y and Chen S 2009 *Phys. Rev. A* **80** 043608
[26] Girardeau M D 2009 *Phys. Rev. Lett.* **102** 245303
[27] Smyrnakis J, Bargi S, Kavoulakis G M, Magiropoulos M, Kärkkäinen K and Reimann S M 2009 *Phys. Rev. Lett.* **103** 100404
[28] Girardeau M D and Astrakharchik G E 2010 *Phys. Rev. A* **81** 043601
[29] Gautam S and Angom D 2010 *J. Phys. B* **43** 095302
[30] Krönke S, Cao L, Vendrell O and Schmelcher P 2013 *New J. Phys.* **15** 063018
[31] Garcia-March M A and Busch T 2013 *Phys. Rev. A* **87** 063633
[32] Anoshkin K, Wu Z and Zaremba E 2013 *Phys. Rev. A* **88** 013609
[33] Cao L, Krönke S, Vendrell O and Schmelcher P 2013 *J. Chem. Phys.* **139** 134103
[34] Peña Ardila L A and Giorgini S 2015 *Phys. Rev. A* **92** 033612
[35] Krönke S, Knrzer J and Schmelcher P 2015 *New J. Phys.* **17** 053001
[36] Petrov D S 2015 *Phys. Rev. Lett.* **115** 155302
[37] Petrov D S and Astrakharchik G E 2016 *Phys. Rev. Lett.* **117** 100401
[38] Schurer J M, Negretti A and Schmelcher P 2017 *Phys. Rev. Lett.* **119** 063001

[39] Cao L, Bolsinger V, Mistakidis S I, Koutentakis G M, Krönke S, Schurer J M and Schmelcher P 2017 *J. Chem. Phys.* **147** 044106
[40] Katsimiga G C, Koutentakis G M, Mistakidis S I, Kevrekidis P G and Schmelcher P 2017 *New J. Phys.* **19** 073004
[41] Michelangeli A and Olgiati A *Anal. Math. Phys.* 2016 http://dx.doi.org/10.1007/s13324-016-0147-3
[42] Klaiman S, Streltsov A I and Alon O E 2017 *Chem. Phys.* **482** 362
[43] Alon O E 2017 *J. Phys. A* **50** 295002
[44] Anapolitanos I, Hott M and Hundertmark D 2017 arXiv:1702.00827v2 [math-ph]
[45] Olgiati A 2017 arXiv:1702.04196v1 [math-ph]
[46] Michelangeli A and Olgiati A 2017 arXiv:1704.00150v1 [math-ph]
[47] Cohen L and Lee C 1985 *J. Math. Phys.* **26** 3105
[48] Yan J 2003 *J. Stat. Phys.* **113** 623
[49] Gajda M 2006 *Phys. Rev. A* **73** 023603
[50] Armstrong J R, Zinner N T, Fedorov D V and Jensen A S 2011 *J. Phys. B* **44** 055303
[51] Lode A U J, Sakmann K, Alon O E, Cederbaum L S and Streltsov A I 2012 *Phys. Rev. A* **86** 063606
[52] Alon O E 2015 *J. Phys.: Conf. Ser.* **594** 012039
[53] Załuska-Kotur M A, Gajda M, Orłowski A and Mostowski J 2000 *Phys. Rev. A* **61** 033613
[54] Schilling C 2013 *Phys. Rev. A* **88** 042105
[55] Benavides-Riveros C L, Toranzo I V and Dehesa J S 2014 *J. Phys. B* **47** 195503
[56] Armstrong J R, Zinner N T, Fedorov D V and Jensen A S 2012 *Phys. Rev. E* **86** 021115
[57] Schilling C and Schilling R 2016 *Phys. Rev. A* **93** 021601
[58] Fasshauer E and Lode A U J 2016 *Phys. Rev. A* **93** 033635
[59] Hall R L 1978 *J. Phys. A* **11** 1227
[60] Hall R L 1978 *J. Phys. A* **11** 1235
[61] Osadchii M S and Muraktanov V V 1991 *Int. J. Quantum Chem.* **39** 173
[62] Armstrong J R, Volosniev A G, Fedorov D V, Jensen A S and Zinner N T 2015 *J. Phys. A* **48** 085301
[63] Bouvrie P A, Majtey A P, Tichy M C, Dehesa J S and Plastino A R 2014 *Eur. Phys. J. D* **68** 346
[64] Cohen-Tannoudji C, Diu B and F. Laloë F 1977 *Quantum Mechanics vol 1* (New York: Wiley)
[65] Vaughan T, Drummond P and Leuchs G 2007 *Phys. Rev. A* **75** 033617
[66] Chen Y J, Pabst S, Li Z, Vendrell O and Santra R 2014 *Phys. Rev. A* **89** 052113
[67] Löwdin P O 1955 *Phys. Rev.* **97** 1474
[68] Coleman A J and Yukalov V I 2000 *Reduced Density Matrices: Coulson's Challenge (Lectures Notes in Chemistry vol 72)* (Berlin: Springer)
[69] Carruthers P and Nieto M M 1968 *Rev. Mod. Phys.* **40** 411
[70] Kechrimparis S and Weigert S 2014 *Phys. Rev. A* **90** 062118
[71] Ma W, Chen B, Liu Y, Wang M, Ye X, Kong F, Shi F, Fei S-M, and Du J 2017 *Phys. Rev. Lett.* **118** 180402

IOP Conf. Series: Journal of Physics: Conf. Series **999** (2018) 012014 doi:10.1088/1742-6596/999/1/012014

Correlated Signals of Free Induction Decay in CdSe–CdS Nanostructures Under Two- and Three-Photon Excitation by Crossed Femtosecond Laser Beams

V V Samartsev[1,2] and T G Mitrofanova[1]

[1]E K Zavoisky Kazan Physical-Technical Institute, Sibirsky Trakt str. 10, Kazan 420029, Russia
[2]Kazan Federal University, Kremlevskaya str. 16, Kazan 420008, Russia

E-mail: samartsev@kfti.knc.ru

Abstract. The possibility and conditions of a generation of the correlated free induction decay signals in CdSe–CdS nanostructures under two- and three-photon excitation by crossed femtosecond beams of the Ti:sapphire laser radiation are considered. It is shown that these correlated signals will be emitted in two opposite directions. The information contained in the temporal shapes and wavefronts of these signals will be correlated.

1. Introduction

Quantum dots are used widely when creating the new lasers, single photon sources and devices for optical quantum computing [1]. Among them an essential role belongs to the colloidal CdSe – CdS quantum dots [2, 3]. In our investigations of various media, including nanostructured samples, we use the various techniques of the coherent optical spectroscopy based on such phenomena as a free induction decay (FID), a photon echo (PE), a four-wave mixing (FWM), etc.

In paper [4] we predicted a possibility to observe the correlated free induction decay (CFID) signals under two-photon excitation of a CdS semiconductor crystal by the femtosecond 800-nm radiation of Ti:sapphire laser in the directions of wave vectors $\mathbf{k}_1$ and $\mathbf{k}_2$ oriented at an angle of 60° to each other. It is known [5] that these signals are formed in the superposition states [6]. Superposition means that there is no possibility to say which of the possibilities, forming the superposition, takes place in reality. Then if the system includes two particles (for example, two electrons), they may be either independent or coupled. Since we consider the coupled particles it should be noted that the predominating mechanism connecting electrons into pairs in our case is the Coulomb interaction. The two-photon femtosecond excitation of coupled electrons in the superposition state leads to the appearance of the nonequilibrium electrical polarization and to the FID generation. But the presence of the nonequilibrium electrical polarization "gratings" $\mathbf{k}_1 - \mathbf{k}_2$ and $\mathbf{k}_2 - \mathbf{k}_1$ causes the simultaneous scattering of two correlated signals CFID1 and CFID2 at once. The phase matching conditions for these signals are

$$\begin{aligned} \mathbf{k}_{\mathrm{CFID1}} &= \mathbf{k}_1 - \mathbf{k}_2, \\ \mathbf{k}_{\mathrm{CFID2}} &= \mathbf{k}_2 - \mathbf{k}_1. \end{aligned} \tag{1}$$

The lifetime of the nonequilibrium electrical polarization "gratings" is defined by the transverse relaxation time T_2 which in the case of the CdS monocrystal at room temperature is equal to 170 fs [4]. The situation is illustrated in figure 1.

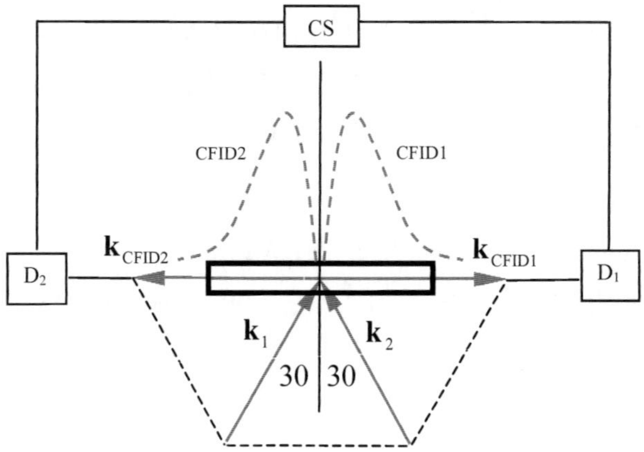

Figure 1. Generation of the correlated signals CFID1 and CFID2 in CdS monocrystal under the two-photon excitation by the 800-nm Ti:sapphire laser radiation in the directions $\mathbf{k}_1$ and $\mathbf{k}_2$. Here $\mathbf{k}_{CFID1}$ and $\mathbf{k}_{CFID2}$ are the wave vectors of the correlated signals CFID1 and CFID2, D_1 and D_2 are the photodetectors, CS is the coincidence scheme.

The novelty of this paper consists in discussion of the possibility to observe such signals in CdSe–CdS nanostructures under the conditions of two- and three-photon excitation. The fact is that unlike a monocrystal the nanostructured samples (such as CdS nanopowder or CdSe–CdS quantum dots) require the three-photon mode for resonant excitation. For the first time the three-photon excitation of nanostructured CdS sample was described by J. Chon and M. Gu in [7] which was then followed by our paper [2]. Here we would like to examine the phase matching conditions for CFID signals under three-photon excitation of CdSe–CdS nanostructures. Besides, the possibility of recording of the correlated transient holograms in such samples will be discussed.

2. Peculiarities of nanostructurded CdSe–CdS samples. Formation of transient "gratings" under two- and three-photon excitation.

As it was noted above, we propose to use the nanostructured CdSe–CdS sample in our experiments. These samples are composites consisting of the CdSe–CdS nanoparticles of "core–shell" type dispersed into a PMMA matrix [3]. The process of construction of such nanoparticles was described in details in [3].

CdSe quantum dots are grown using the colloidal synthesis technique. They are then covered with the CdS shell. The average size of the constructed nanoparticles is equal to $3{\cdot}10^{-7}$ cm. They have either spherical or ellipsoidal shape. The Ti:sapphire laser radiation has the wavelength of $8{\cdot}10^{-5}$ cm with pulse duration of 40 fs. The area of laser excitation according to [3] is about $3{\cdot}10^{-5}$ cm. In this case the number of working particles is estimated as $3{\cdot}10^{4}$. The coherence length $c\tau_0$ (where c is the light velocity and τ_0 is the coherence time which is supposed to be equal to $T_2 = 170$ fs) is equal to 0.5 cm. The size of the sample is limited by this value which can be increased up to 1 cm.

IOP Conf. Series: Journal of Physics: Conf. Series **999** (2018) 012014 doi:10.1088/1742-6596/999/1/012014

It should be noted that the luminescence spectrum of the CdSe semiconductor having the forbidden zone in the IR range lies in the visible diapason (for the nanoparticles of size ≤ 3 nm). The quantum emission is high enough since electrons and holes are localized in nanometer volume inside the quantum dot [8]. In our case of excitation by short femtosecond pulses the connected excitons can play the role of qubit.

Depending on the nanostructure size the bandgap width varies and the quantum size effect takes place which reveals in changing of the luminescence wavelength [2]. The CdSe–CdS nanoparticles of 2, 3, 6.5 and 7 nm size have a maximum of the emission line at 525, 540, 590 and 600 nm, respectively. It was established [2] that covering the CdSe core with the CdS shell allows an eight-fold increase in luminescence. The fact is that such heterostructures have the special configuration of the energy zones. A smaller number of internal elementary excitations decays nonradiatively [2]. That is why the excitation becomes more effective. Under the excitation by the Ti:sapphire laser radiation CdSe–CdS nanoparticles demonstrate the nonlinear properties such as multi-photon absorption and luminescence at the above mentioned wavelengths. The CdSe–CdS band structure is shown in figure 2.

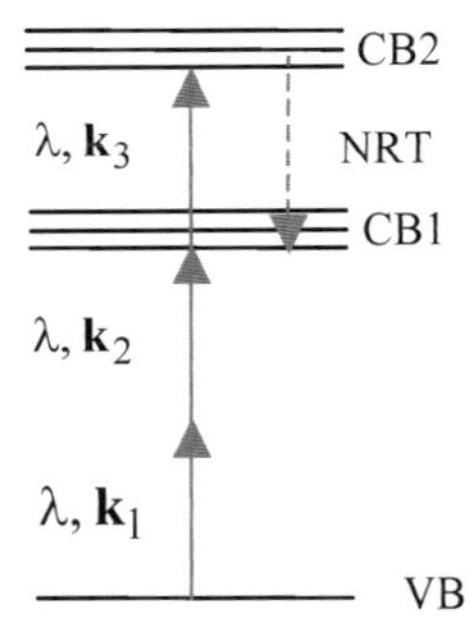

Figure 2. The CdSe–CdS band structure and the scheme of energy transitions under a three-photon excitation by the Ti:sapphire laser radiation: VB – valence band, CB1 and CB2 – conductive bands, NRT – nonradiative transition.

Let us now discuss the three-photon mode of excitation of the nanostructured CdSe–CdS sample. After excitation by three femtosecond pulses of the same wavelength (800 nm) the electrons of the sample are transferred to the upper conductive band CB2 corresponding to the wavelength of 267 nm (i.e. $\lambda_{exc} = \lambda/3$). Then according to the experimental results of paper [9], the electrons decay via nonradiative transition to the conductive band CB1 corresponding to the wavelength of 534 nm (i.e. 267 nm × 2) and then decay radiatively. Notice that at the same time an independent two-photon transition takes place between the valence VB and conductive CB1 bands and the nonequilibrium electrical polarization "gratings" described above are formed and then scatter the CFID1 and CFID2 signals.

We would like to write the phase matching conditions for these signals under three-photon mode of excitation. If there are two exciting femtosecond beams then the third exciting pulse can act only in the directions of wave vectors $\mathbf{k}_1$ or $\mathbf{k}_2$. Suppose that the nonradiative transition is absent. In this case, the phase matching conditions could be written as:

$$\mathbf{k}_{\mathrm{CFID1}} = 2\mathbf{k}_1 - \mathbf{k}_2, \qquad (2)$$
$$\mathbf{k}_{\mathrm{CFID2}} = 2\mathbf{k}_2 - \mathbf{k}_1.$$

But in the case of CdSe–CdS due to the nonradiative transition CB2→CB1 the information about the phase of the third exciting pulse and about the population of levels belonging to the conductive band CB2 is totally lost. The situation seems to be hopeless but there is the independent energy transition VZ→CB1 and thanks to this the transient gratings $\mathbf{k}_1 - \mathbf{k}_2$ and $\mathbf{k}_2 - \mathbf{k}_1$ are induced and the correlated responses CFID1 and CFID2 will be emitted.

In conclusion, let us say some words about the possibility of the femtosecond holograms recording under a multi-photon regime of absorption of laser radiation. For the first time, such possibility have been shown experimentally by A.K. Rebane with colleagues in an organic sample [10]. Then, in our paper [11], the formation of the correlated transient holograms (CTH) in the CdS crystal was considered for the case of the two-photon mode of an excitation. We suggested recording of the holograms into the wave fronts of signals CFID1 and CFID2 in the experimental scheme presented in figure 1. It was found that these holograms are reversed and for the spherical wave fronts the following expressions were obtained:

$$\begin{aligned} \frac{1}{R_{\mathrm{CTH1}}} &= \frac{1}{R_1} - \frac{1}{R_2}, \\ \frac{1}{R_{\mathrm{CTH2}}} &= \frac{1}{R_2} - \frac{1}{R_1}, \end{aligned} \tag{3}$$

where R_1 and R_2 are the radius of curvature of the corresponding exciting waves. Since CFID signals can be also observed under three-photon mode of CdSe–CdS excitation they can be used for the recording of the correlated transient holograms in such nanostructures.

Acknowledgments

This study was supported by RAS Program "Actual problems of physics of low temperatures" and RFBR grant no. 17-02-00701-a.

References

[1] Bera D, Qian L, Tseng T-K and Holloway P 2010 Materials **3** 2260
[2] Zharkov D K, Shmelev A G, Leontiev A V, Lobkov V S and Samartsev V V in *Coherent optics and optical spectroscopy* (Salakhov M Kh, ed.) (Kazan: Kazan Federal University) 2014, p.109-112 (in Russian)
[3] Leontiev A V, Zharkov D K, Shmelev A G, Nikiforov V G, Lobkov V S in *Coherent optics and optical spectroscopy* (Salakhov M Kh, ed.) (Kazan: Kazan Federal University) 2015, p.146-148 (in Russian)
[4] Leontiev A V, Lobkov V S, Mitrofanova T G, Samartsev V V and Shmelev A G 2012 *Laser Phys. Lett.* **9** 654
[5] Samartsev V V 2015 *Correlated Photons and Their Applications* (Cambridge, UK: CISP)
[6] Yevseyev I V, Rubtsova N N and Samartsev V V 2009 *Coherent Transient Processes in Optics* (Moscow: Fizmatlit) (in Russian)
[7] Chon J W M and Gu M 2004 *Appl. Phys. Lett.* **84** 4472
[8] Novotny L, Hecht B 2006 *Principles of nano-optics* (Cambridge: Univ. Press)
[9] Zharkov D K, Safiullin G M, Nikiforov V G, Lobkov V S and Samartsev V V 2013 *Uchenye Zapiski Kazanskogo Universiteta. Seriya Fiziko-Matematicheskie Nauki* **155** 66 (in Russian)
[10] Rebane A, Drobizhev M and Karotki A 2002 *Jour. of Luminescence* **98** 341
[11] Samartsev V V, Leontiev A V and Mitrofanova T.G. 2016 *Jour. of Physics: Conf. Ser.* **714** 012014

IOP Conf. Series: Journal of Physics: Conf. Series **999** (2018) 012015 doi:10.1088/1742-6596/999/1/012015

Dynamics of Nonlinear Excitation of the High-Order Mode in a Single-Mode Step-Index Optical Fiber

V Burdin[1] and A Bourdine[1]

[1]Povolzhskiy State University of Telecommunications and Informatics (PSUTI), 23, Lev Tolstoy str., 443010, Samara, Russia

E-mail: burdin@psati.ru

Abstract. This work is concerned with approximate model of higher-order mode nonlinear excitation in a singlemode silica optical fiber. We present some results of simulation for step-index optical fiber under femtosecond optical pulse launching, which confirm ability of relatively stable higher-order mode excitation in such singlemode optical fiber over sufficiently narrow range of launched optical power variation.

1. Introduction

It is well known that optical fibers may operate in a single-mode, multi-mode or few-mode regimes. It depends on ratio between wavelength and optical fiber core diameter as well as optical emission launching conditions and refractive index profile height and form [1 – 5]. For example, in singlemode optical fibers higher-order modes do not satisfy cut-off condition over corresponding operation wavelength band, and here only one fundamental mode is supported under linear propagation conditions, while dozens of guided modes can propagate in large core multimode optical fibers since they satisfy cut-off condition over operation wavelength band. However when laser-based optical emission is launched to the core of multimode optical fiber, only a limited number of modes from the total quality is excited, and optical fiber operates in a few-mode regime.

During high power optical emission propagates over optical fiber, its refractive index profile form is able to change due to nonlinearity factors [6, 7]. In fact this leads to changing of launching conditions and new guided more higher-order modes can be excited and supported by optical fiber in comparison with linear regime [8, 9]. Moreover even for singlemode optical fiber a higher-order mode can satisfy cut-off condition in operation wavelength under nonlinear excitation.

Recently usually nonlinear mode excitation was researched for analysis of fiber optic amplifier mode instability, where optical fiber refractive index profile changes due to heating effects. However the great achievements in the field of femtosecond laser technique permits to consider ability of mode staff management by Kerr nonlinearity [8, 12]. In previous works [13, 14] we demonstrated ability of formulation of necessary conditions for higher-order mode LP_{11} excitation in silica step-index singlemode optical fiber by nonlinear Kerr effect before irreversible changes of optical fiber refractive index would start up to its destruction. However cut-off condition is just necessary but no sufficient for nonlinear mode excitation. Research of this effect requires analysis of process dynamics. This work presents simulation results of nonlinear higher-order mode excitation process dynamics in silica step-index singlemode optical fiber.

2. Generalized solution for numerical simulation of optical pulse propagation along singlemode and multimode optical fibers

Femtosecond optical pulse propagation along multimode optical fiber is usually described by system of coupled nonlinear Schrödinger equations (CNLSE), each of them can be represented in the following form [6, 15 – 16]:

$$\frac{\partial A_m}{\partial z} + \beta_{m1}\frac{\partial A_m}{\partial \tau} - j\frac{\beta_{m2}}{2}\frac{\partial^2 A_m}{\partial \tau^2} + \frac{\beta_{m3}}{6}\frac{\partial^3 A_m}{\partial \tau^3} + \frac{\alpha_m}{2}A_m - j\gamma\cdot\sum\left[C_{n,m}\left|A_n^2\right|A_m + \right.$$
$$\left. + j\frac{B_{n,m}}{\omega_0}\frac{\partial\left(\left|A_n^2\right|A_m\right)}{\partial \tau} - T_R B_{n,m}\frac{\partial\left|A_n^2\right|}{\partial \tau}A_m\right] = 0, \tag{1}$$

where A_m is optical pulse spatial envelope transferred by m-th mode in optical fiber; β_{m1} is m-th mode delay parameter; β_{m2} is m-th mode chromatic dispersion parameter; β_{m3} is m-th mode second-order chromatic dispersion parameter; α_m is m-th mode attenuation parameter; $C_{n,m}$ and $B_{n,m}$ are mode coupling coefficients; ω_0 is the carrier frequency; T_R is Raman time constant; τ is normalized time; z is the distance.

In general equation system (1) is widely used for admissible simulation of optical pulse propagation with the pulse width more 10 fs. However under carrier frequency ω_0 corresponding to wavelength λ=800 nm typical Raman time constant is T_R=5 fs [6, 15 – 17], while for 10-fs and wider optical pulses the equation system (1) is simplified that the last two terms of sum can be neglected [6, 15 – 18].

Usually for numerical simulation of optical pulse nonlinear propagation the equation system (1) is solved by well known Split-Step Fourier Method (SSFM) [6, 15 – 18]. Here CNLSE (1) is rewritten in the following form:

$$\frac{\partial A_m}{\partial z} = \left(\hat{D} + \hat{N}\right)A_{\mathrm{m}}, \tag{2}$$

where $\hat{D}$ and $\hat{N}$ are linear and nonlinear operators taking into account linear and nonlinear optical pulse distortions. Symmetric split-step scheme is usually used with second-order accuracy over integration step h [15 – 18]. Solution of (2) is written in the following form:

$$A_m(z+h,\tau) \approx \exp\left(\frac{h}{2}\hat{D}\right)\exp\left(h\hat{N}\right)\exp\left(\frac{h}{2}\hat{D}\right)A_m(z,\tau), \tag{3}$$

Linear part $\exp\left(h\hat{D}/2\right)$ is performed in frequency space. It seems clear that the number of equations in CNLSE is defined by the number of modes propagating over researched optical fiber.

3. Higher-order mode nonlinear excitation conditions for singlemode optical fiber

Singlemode optical fibers support only one fundamental mode. For weakly guiding optical fibers it is the mode LP_{01}. Therefore usually under low power of transmitted optical emission that corresponds to linear regime, singlemode optical fiber provides propagation for only one fundamental mode over operation wavelength band. In this case optical signal is transferred by two orthogonal polarization modes which degenerate to one fundamental mode LP_{01} under ideal symmetry of weakly guiding optical fiber.

In works [13, 14] it was demonstrated that under sufficiently great power level of optical emission more than some threshold value P_{th}, when refractive index profile changing due to Kerr nonlinearity should be taken into account and can not be neglected, necessary conditions for higher-order mode LP_{11} excitation in singlemode step-index optical fibers are provided. Moreover for femtosecond pulses this power level is less than critical power levels Pcr, when refractive index irreversibly changes occur. It was shown in works [13, 14] the threshold power depends on normalized frequency linearly.

4. Model of pulse propagation along singlemode step-index optical fiber under taking into account higher-order mode nonlinear excitation

In this work simulation of higher-order mode nonlinear excitation dynamics in singlemode optical fiber is based on CNLSE (1) solution by SSFM. Here we utilize parameter

$$\overline{P_L} = \left|A_{01}^2\right| + \left|A_{11}^2\right|, \tag{4}$$

where A_{01} и A_{11} are envelopes of optical pulse mode components LP_{01} and LP_{11}. It is supposed that under $P_L < P_{th}$ optical pulse propagation process is described by only one equation written in relation to envelope A_{01} for the fundamental mode. $P_L > P_{th}$ corresponds to conditions for mode LP_{11} excitation, and pulse propagation is represented by two CNLSEs for modes LP_{01} and LP_{11}, while during $P_L > P_{cr}$ optical pulse does not propagate and $A_{01}=0$, $A_{11}=0$ due to optical fiber destruction under such great values of optical power.

During simulation we use algorithm proposed in works [20 – 23] that involves to symmetric split-step scheme of SSFM an additional analysis of researched optical fiber with new refractive index profile deformed due to nonlinearity before nonlinear operator execution. Here parameters of modes LP_{01} and LP_{11} are computed by Gaussian approximation. We utilized analytical expressions derived for those modes by taking into account Kerr nonlinearity and represented in works [24 – 26] and developed following model described by following expressions:

$$\frac{\partial A_{01}}{\partial z}\hat{D}_0 A_{01} + \hat{N}_{D_0} A_{01} + \hat{N}_{N_1} A_{01} = 0;\ \ \overline{P}_L < P_{th}; \tag{5}$$

$$\begin{cases} \dfrac{\partial A_{01}}{\partial z} + \hat{D}_{01}A_{01} + \hat{N}_{D,01}A_{01} + \hat{N}_{N_2}\left(A_{01}, A_{01}\right) + \hat{N}_{N_2}\left(A_{01}, A_{11}\right) = 0 \\ \dfrac{\partial A_{11}}{\partial z} + \hat{D}_{11}A_{11} + \hat{N}_{D,11}A_{01} + \hat{N}_{N_2}\left(A_{11}, A_{11}\right) + \hat{N}_{N_2}\left(A_{11}, A_{01}\right) = 0 \end{cases};\ \ \overline{P}_L > P_{th}; \tag{6}$$

$$\begin{cases} A_{01} = 0 \\ A_{11} = 0 \end{cases};\ \ \overline{P}_L > P_{cr}; \tag{7}$$

where

$$\hat{D}A_{01} = \beta_{01,1}\frac{\partial A_{01}}{\partial \tau} - j\frac{\beta_{01,2}}{2}\frac{\partial^2 A_{01}}{\partial \tau^2} + \frac{\beta_{01,3}}{6}\frac{\partial^3 A_{01}}{\partial \tau^3} + \frac{\alpha_{01}}{2}A_{01};$$

$$\hat{D}A_{11} = \beta_{11,1}\frac{\partial A_{11}}{\partial \tau} - j\frac{\beta_{11,2}}{2}\frac{\partial^2 A_{11}}{\partial \tau^2} + \frac{\beta_{11,3}}{6}\frac{\partial^3 A_{11}}{\partial \tau^3} + \frac{\alpha_{11}}{2}A_{11};$$

$$\hat{N}_{01}A_{01} = j\gamma \cdot \left[\left|A_{01}^2\right| + j\frac{1}{\omega_0 A_{01}}\frac{\partial\left(\left|A_{01}^2\right|A_{01}\right)}{\partial \tau} - T_R\frac{\partial\left|A_{01}^2\right|}{\partial \tau}\right]\frac{A_{01}}{4R_{01}^2};$$

$$\hat{N}_{11}A_{11} = j\gamma \cdot \left[\left|A_{11}^2\right| + j\frac{1}{\omega_0 A_{01}} \frac{\partial\left(\left|A_{11}^2\right|A_{01}\right)}{\partial\tau} - T_R \frac{\partial\left|A_{11}^2\right|}{\partial\tau} \right] \frac{A_{01}}{4R_{11}^2};$$

$$\hat{N}_{01}A_{11} = j\gamma \cdot \left[\left|A_{01}^2\right| + j\frac{1}{\omega_0 A_{11}} \frac{\partial\left(\left|A_{01}^2\right|A_{11}\right)}{\partial\tau} - T_R \frac{\partial\left|A_{01}^2\right|}{\partial\tau} \right] \frac{A_{11}}{4R_{01}^2};$$

$$\hat{N}_{11}A_{11} = j\gamma \cdot \left[\left|A_{11}^2\right| + j\frac{1}{\omega_0 A_{11}} \frac{\partial\left(\left|A_{11}^2\right|A_{11}\right)}{\partial\tau} - T_R \frac{\partial\left|A_{11}^2\right|}{\partial\tau} \right] \frac{A_{11}}{4R_{11}^2}.$$

Here we take into account that mode field radiuses are twice as much as equivalent radiuses of Laguerre-Gaussian modes R_{01} и R_{11} [27]. Linear operators are performed in frequency space by fast direct and inverse Fourier transformation.

5. Results of simulation of higher-order mode nonlinear excitation dynamics in a singlemode optical fiber

We performed simulation of LP_{11} mode nonlinear excitation for step-index silica optical fiber with numerical aperture NA=0.16, GeO_2-doped core with diameter 8.3 μm and pure SiO_2 outer solid cladding. Total researched optical fiber length is 10 mm. We considered 12-fs optical pulse generating by laser source operating at wavelength λ=800 nm injecting to optical fiber core under overfilled launching conditions. Power loss occurring during optical signal propagation over optical fiber was neglected. Example of LP_{01} and LP_{11} mode evolution is represents on Fig. 1.

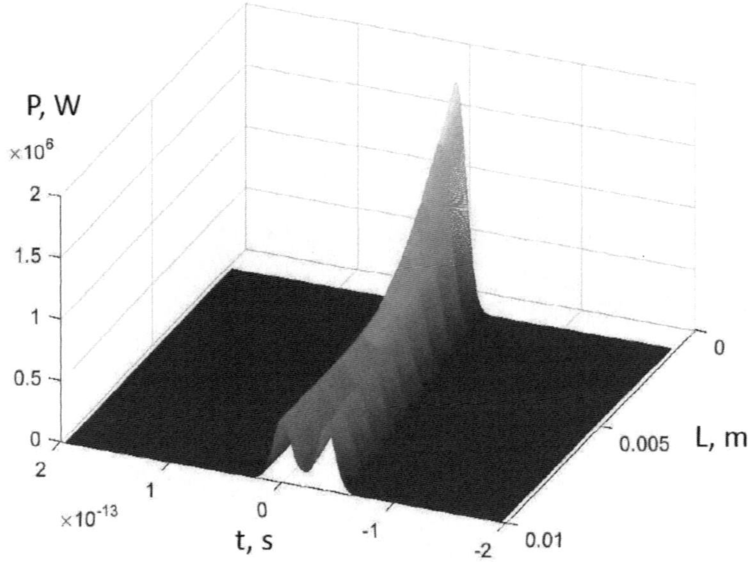

Figure 1. LP_{01} and LP_{11} mode evolution in singlemode optical fiber

Here injected power was 1.1 MW under threshold power 1.0 MW, and we have got stable LP_{11} mode component under described conditions. However this effect was achieved in sufficiently narrow range of optical power variations. Thus during injected optical power reducing LP_{11} mode nonlinear excitation conditions were violated, while optical emission intensity quickly achieved threshold value under following improvement of launched optical power. Determination of allowed optical power value range for stable higher-order mode nonlinear excitation requires more detailed researching. Represented results undoubtedly are preliminary due to proposed model is approximate, and it need experimental verification. However computed results of simulation provide to suppose ability of stable higher-order mode nonlinear excitation in a singlemode optical fiber.

6. Conclusion

We described approximate model for simulation of higher-order mode nonlinear excitation in a singlemode step-index optical fiber and presented some results of this process dynamics computation. In spite of simulation results are preliminary, they demonstrate ability of stable higher-order mode nonlinear excitation in a singlemode optical fiber during femtosecond optical pulse transmission over sufficiently narrow range of injected optical power variation.

7. References

[1] Katzenelenbaum B Z 1961 *Theory of Irregular Waveguides with Slow Changing Parameters* (Moscow: Academy of Sciences of the USSR)

[2] Snyder A W and Love J 1983 *Optical Waveguide Theory* (London: Chapman & Hall)

[3] Black R J and Gagnon L 2010 *Optical Waveguide Modes: Polarization, Coupling and Symmetry* (NY: McGraw-Hill Education)

[4] Raevskii A S and Raevskii S B 2010 *Complex Waves* (Moscow: Radiotekhnika)

[5] Andreev V A and Bourdine A V 2004 *Multimode Optical Fibers: Theory and applications for high bit rate networks* (Moscow: Radio i Svyaz)

[6] Agrawal G P 2001 *Nonlinear fiber optics* (Amsterdam: Elsevier – Academic Press)

[7] Lancry M, Poumellec B, Chahid-Erraji A, Beresna M and Kazansky P G 2011 Dependence of the femtosecond laser refractive index change thresholds on the chemical composition of doped-silica glasses *Opt. Mat. Expr.* **1**(4) 711 – 723

[8] Romanova E A, Melnikov L A and Bekker E V 2001 Light guiding in optical fibers with Kerr-like nonlinearity *Micr. and Opt. Tech. Letts.* 2001 **30**(3) 212 – 216

[9] Hansen K R, Alkeskjold T T, Broeng J and Laegsgaard J 2013 Theoretical analysis of mode instability in high-power fiber amplifiers *Opt. Express* **21**(2) 1944 – 71

[10] Smith A V and Smith J J 2013 Review of models of mode instability in fiber amplifiers *As-Photonics Publ.* 2013

[11] Dianov E M, Zakhidov E A, Karasik A Ya, Marnyshev P V and Prokhorov A M 1982 Stimulated four-photon nonlinear processes in few-mode glass-fiber optic waveguides *Sov. Phys. JETP* **56**(1) 21 – 26

[12] Liu Z, Wright L G, Christodoulides D N and Wise F W 2016 Kerr self-cleaning of femtosecond-pulsed beams in graded-index multimode fiber *Opt. Lett.* **41**(16) 3675 – 78

[13] Burdin V A, Bourdine A V and Dmitriev E V 2017 Necessary conditions for nonlinear excitation of a high-order mode in a single-mode optical fiber *Opt. and Spectr.* **123**(2) 193 – 199

[14] Burdin V A and Bourdine A V 2017 Necessary conditions for a nonlinear few-mode excitation of modes in single-mode optical fiber *Proc. of SPIE* **10342** 10342-0M – 10M

[15] Agrawal G P 2011 *Applications of Nonlinear Fiber Optics* (NY: Academic Press)

[16] Deiterding R, Glowinski R, Oliver H and Poole S 2013 A reliable split-step Fourier method for the propagation equation of ultrafast pulses in single-mode optical fibers *J. Light. Tech.* **31**, 2008 –17

[17] Deiterding R and Poole S W 2016 Robust split-step Fourier methods for simulating the propagation of ultra-short pulses in single- and two-mode optical communication fibers *Splitting Methods in Communication, Imaging, Science, and Engineering. Scientific Computation* ed R Glowinski, S Osher, W Yin (Cham: Springer) chapter 18 pp 603 – 626

[18] Nakamura S 2010 Comparison between Finite-Difference Time-Domain method and experimental results for femtosecond laser pulse propagation, coherence and ultrashort pulse laser emission *Coherence and Ultrashort Pulse Laser Emission* ed F J Duarte (InTech) chapter 20 449 – 492

[19] Fedoruk M P and Sidelnikov O S 2015 Algorithms for numerical simulation of optical communication links based on multimode fiber *Computational Technologies* **20**(5) 105 – 119

[20] Burdin V A and Bourdine A V 2017 Dispersion characteristics of step index single mode optical fiber with Kerr nonlinearity *Proc. of SPIE* **10342** 10342-0N – 11N
[21] Andreev V A, Bourdine A V, and Burdin V A and Kubanov V P 2017 Solution for the fundamental mode of a step-index lightguide with Kerr nonlinearity *Infocom. Tech.* **15**(4) 313 – 320
[22] Bourdine A V and Burdin V A 2017 Dispersion characteristics of the LP11 mode of a step-index optical fiber with Kerr nonlineatity *Proceedings of Educational Communication Institutions* **3**(4) 5 – 9
[23] Petermann K 1977 Fundamental mode microbending loss in graded-index and W-fibres *Opt. and Quant. Electr.* **9** 167 – 175

Acknowledgments
The reported study was funded by RFBR according to the research project No. 16-37-60015 mol_a_dk

Channel Capacity Calculation at Large SNR and Small Dispersion within Path-Integral Approach

A V Reznichenko[1,2], **I S Terekhov**[1,2]

[1] Theory department, Budker Institute of Nuclear Physics of Siberian Branch Russian Academy of Sciences, Novosibirsk 630090, Russia
[2] Novosibirsk State University, Novosibirsk 630090, Russia

E-mail: a.v.reznichenko@inp.nsk.su

Abstract. We consider the optical fiber channel modelled by the nonlinear Shrödinger equation with additive white Gaussian noise. Using Feynman path-integral approach for the model with small dispersion we find the first nonzero corrections to the conditional probability density function and the channel capacity estimations at large signal-to-noise ratio. We demonstrate that the correction to the channel capacity in small dimensionless dispersion parameter is quadratic and positive therefore increasing the earlier calculated capacity for a nondispersive nonlinear optical fiber channel in the intermediate power region. Also for small dispersion case we find the analytical expressions for simple correlators of the output signals in our noisy channel.

1. Introduction

We consider the fiber optical channel described by the nonlinear Schrödinger equation (NLSE) with additive white Gaussian noise for the case when the signal-to-noise ratio (SNR) is large:

$$\partial_z \psi + i\beta\partial_t^2 \psi - i\gamma|\psi|^2\psi = \eta(z,t)\,, \tag{1}$$

here $\psi(z,t)$ is the complex signal, γ is the Kerr nonlinerity, $\beta = \beta_2/2$ is the dispersion parameter, $\eta(z,t)$ is the white Gaussian noise: $\langle\eta(z,t)\rangle_\eta = 0$, $\langle\eta(z,t)\bar{\eta}(z',t')\rangle_\eta = Q\delta(z-z')\delta(t-t')$, where Q is a power of the white Gaussian noise (per unit length and frequency). The bar here and hereafter means complex conjugation.

We consider the mutual information [1] of the channel for the given input (continuous) signal probability density function (PDF) $P[X]$: $\psi(z=0,t) = X(t)$. In our approach we use the representation of the mutual information in the path-integral form

$$I_{P[X]} = \int \mathcal{D}X\mathcal{D}Y P[X]P[Y|X]\log\left[\frac{P[Y|X]}{P_{out}[Y]}\right] = H[Y] - H[Y|X], \tag{2}$$

and we assume that the input signal X has the fixed finite average power P_{ave}. The function $P[Y|X]$ in (2) is the conditional probability density function, that is the probability density of receiving output signal $\psi(z=L,t) = Y(t)$ when the input signal is given. The output signal PDF reads $P_{out}[Y] = \int \mathcal{D}X P[X]P[Y|X]$. The quantity $H[Y] = -\int \mathcal{D}Y P_{out}[Y]\log\left[P_{out}[Y]\right]$ is the output signal entropy (responsible for the signal transmission), and the entropy

$H[Y|X] = -\int \mathcal{D}X\mathcal{D}Y P[X]P[Y|X]\log\left[P[Y|X]\right]$ is the conditional entropy responsible for the noise impact,i.e., it is the measure of the uncertainty of the output signal $Y(t)$ for a given input signal $X(t)$. The measures $\mathcal{D}X$ and $\mathcal{D}Y$ in (2) are defined below.

In our approach we use the formulation for the conditional PDF $P[Y|X]$ through the path-integral (Martin-Siggia-Rose formalism [2]). In Ref.[3] the following ("quasiclassical") representation for the conditional probability density function was obtained:

$$P[Y|X] = \Lambda\exp\left\{-\frac{S[\Psi_{cl}]}{Q}\right\} \approx e^{-\frac{\Gamma[\Psi_{cl}]}{Q}}, \quad \Lambda = \int\limits_{\phi(z=0)=0}^{\phi(z=L)=0} \mathcal{D}\phi\exp\left\{-\frac{S[\Psi_{cl}+\phi]-S[\Psi_{cl}]}{Q}\right\}, \tag{3}$$

where the action $S[\psi]$ in (3) reads in the time domain as quadrated left hand side of NLSE (1):

$$S[\psi] = \int\limits_0^L dz \int\limits_T dt\,|\mathcal{L}[\psi(z,t)]|^2\,, \quad \mathcal{L}[\psi] = \partial_z\psi(z,t) + i\beta\partial_t^2\psi(z,t) - i\gamma\psi(z,t)|\psi(z,t)|^2\,. \tag{4}$$

The approximate ("saddle point" approximation) equation in Eq. (3) involves the one-loop theoretical field effective action $\Gamma[\Psi_{cl}]$ corresponding to the "classical action" (4), see [4]. Our goal is to calculate both the action $S[\Psi_{cl}]$ and the normalization factor Λ.

We introduce T as a time interval containing both signals $X(t)$ and $Y(t)$. In what follows we use the discretization scheme in the time domain. This scheme uses the dense time grid with M' intervals and spacing δ_t (i.e., $t_j = -T/2 + j\delta_t$, $j = 0, 1, \ldots, M'-1$) and the coarse time sub-grid with M intervals and spacing $\widetilde{\delta}_t$: $T = M'\delta_t = M\widetilde{\delta}_t$, $\delta_t \propto 2\pi/W'$, $\widetilde{\delta}_t \propto 2\pi/W$. Here W is the bandwidth of the input signal $X(t_j) = \rho(t_j)\exp[i\phi_0(t_j)]$, and W' is the noise bandwidth (usually, $W' \gg W$). The measure in (3) in our scheme reads

$$\mathcal{D}\phi(z,t) = \lim_{\delta_t\to 0}\lim_{\Delta\to 0}\left(\frac{\delta_t}{\Delta\pi Q}\right)^{M'}\prod_{j=0}^{M'-1}\prod_{i=1}^{N-1}\left\{\frac{\delta_t}{\Delta\pi Q}\, dRe\phi(z_i,t_j)\, dIm\phi(z_i,t_j)\right\}, \tag{5}$$

where $\Delta = L/N$ is the coordinate z grid spacing with N points on L, L is the channel length.

The function Ψ_{cl} in Eq. (3) is the solution (referred to as "the classical solution") of the Euler-Lagrange equation $\delta S[\Psi_{cl}] = 0$ with the boundary conditions: $\Psi_{cl}(z=0) = X$, $\Psi_{cl}(z=L) = Y$. In the time domain this equation for $\Psi_{cl}(z,t)$ has a notedly simple form [5]

$$\begin{aligned}&\left(\partial_z + i\beta\partial_t^2 - 2i\gamma|\Psi_{cl}(z,t)|^2\right)\mathcal{L}[\Psi_{cl}(z,t)] + i\gamma\Psi_{cl}^2(z,t)\overline{\mathcal{L}[\Psi_{cl}(z,t)]} = 0,\\ &\mathcal{L}[\Psi_{cl}(z,t)] = \left(\partial_z + i\beta\partial_t^2 - i\gamma|\Psi_{cl}(z,t)|^2\right)\Psi_{cl}(z,t).\end{aligned} \tag{6}$$

It is convenient to introduce the function $\Phi(z,t)$ which is the solution of the nonlinear Schrödinger equation (NLSE) with zero noise, i.e., $\mathcal{L}[\Phi(z,t)] = 0$, and with the boundary condition $\Phi(0,t) = X(t)$. It is obvious that the function $\Phi(z,t)$ obeys the equation Eq. (6) and the boundary condition at $z = 0$, but it does not obey the boundary condition at $z = L$. It globally minimizes the action as well: $S[\Phi(z,t)] = 0$. Since we imply that the noise power is much less than the signal power we can present the solution of Eq. (6) in the form $\Psi_{cl}(z,t) = \Phi(z,t) + \varkappa(z,t)$, where the function $\varkappa(z,t)$ is of order of $\sqrt{Q}$ for unsuppressed configurations $\Psi(z,t)$, since in the leading order in 1/SNR the action $S[\Psi_{cl}]$ is quadratic functional in $\varkappa(z,t)$, see details in [5]. Therefore we substitute the function $\Psi_{cl}(z,t)$ in this form to the Eq. (6), then linearizing Eq. (6) in $\varkappa(z,t)$ we obtain the *linear problem* on $\varkappa(z,t)$:

$$\begin{aligned}&\left(\partial_z + i\beta\partial_t^2 - 2i\gamma|\Phi(z,t)|^2\right)l[\varkappa(z,t)] + i\gamma\Phi^2(z,t)\overline{l[\varkappa(z,t)]} = 0,\\ &l[\varkappa(z,t)] = \left(\partial_z + i\beta\partial_t^2\right)\varkappa(z,t) - i\gamma\left(2\varkappa(z,t)|\Phi(z,t)|^2 + \overline{\varkappa(z,t)}\Phi^2(z,t)\right),\end{aligned} \tag{7}$$

with the boundary conditions $\varkappa(z=0,t)=0,\qquad \varkappa(z=L,t)=Y(t)-\Phi(L,t)\equiv\delta Y(t)$.

We will use the dimensionless dispersion parameter $\widetilde{\beta}=\beta L/\delta_t^2$. In our perturbative consideration this parameter is assumed to be small: $\widetilde{\beta}\ll 1$. The dimensionless parameter $\widetilde{\gamma}=\gamma L P_{ave}/\sqrt{3}$ is the average nonlinearity parameter, where P_{ave} is the average power over the input signal PDF $P[X]$: $P_{ave}=\lim_{T\to\infty}\int\mathcal{D}X\,P[X(t)]\frac{1}{T}\int_T|X(t)|^2dt$.

2. Conditional PDF $P[Y|X]$ for small dispersion parameter

In the leading order in 1/SNR the action difference in (3) reads in discrete time form [5]:

$$S[\Psi_{cl}+\phi]-S[\Psi_{cl}]\approx S_2[\phi,\Phi]=\delta_t\sum_{k=0}^{M'-1}\int_0^L dz\,\mathcal{L}_{eff}[\phi(z,t_k)],\tag{8}$$

$$\mathcal{L}_{eff}[\phi(z,t_k)]=\left|\partial_z\phi(z,t_k)+i\beta\partial_t^2\phi(z,t_k)-i\gamma\left(2\phi(z,t_k)|\Phi(z,t_k)|^2+\bar{\phi}(z,t_k)\Phi^2(z,t_k)\right)\right|^2.\tag{9}$$

We perform our calculation in the standard discretisation scheme where the second time derivative in the path-integral (3) is replaced with the second order discrete analog:

$$\partial_t^2\psi(z,t_i)\longrightarrow\Delta_{\delta t}^2\psi(z,t_i)=\frac{\psi(z,t_i+\delta t)-2\psi(z,t_i)+\psi(z,t_i-\delta t)}{\delta t^2}.\tag{10}$$

Note that different discretisation schemes of the time second derivative correspond to the variation of time grid spacing (i.e., different renormalisation schemes). For instance, for the second difference derivation $\Delta_{(3)}^2$ of the third order approximation we have $\Delta_{(3)}^2\psi(z,t_i)\equiv -\psi(z,t_i+2\delta t)+16\psi(z,t_i+\delta t)-30\psi(z,t_i)+16\psi(z,t_i-\delta t)-\psi(z,t_i-2\delta t)/(12\delta t^2)=\left(\frac{4}{3}\Delta_{\delta t}^2-\frac{1}{3}\Delta_{2\delta t}^2\right)\psi(z,t_i)$, i.e., it is equivalent to the consideration of two combined grids δt and $2\delta t$ with the same operator (10). In this sense, our fixed grig spacing $\delta t\propto 2\pi/W'$ is concerned with the bandwidth W' of the noise, and it places the role of the ultraviolet cutoff of our field theory with the action $S[\psi]$, see (4).

Based on results of [6] for the conditional PDF $P[Y|X]$ of a nondispersive nonlinear optical fiber channel we can present the following result for M' independent time channels governed by NLSE (1) with zero noise and $\beta=0$:

$$P[Y|X]^{(\beta=0)}=\exp\left\{-\delta_t\sum_{j=0}^{M'-1}\frac{(1+4\mu_j^2/3)x_{0,j}^2-2\mu_j x_{0,j}y_{0,j}+y_{0,j}^2}{QL(1+\mu_j^2/3)}\right\}\prod_{j=0}^{M'-1}\frac{\delta_t}{\pi QL\sqrt{1+\mu_j^2/3}}.\tag{11}$$

Here $\mu(t_j)=\mu_j=\gamma L\,\rho^2(t_j)$, and real functions $x_{0,j}=x_0(t_j)$ and $y_{0,j}=y_0(t_j)$ are defined through the output signal $Y(t)$ and the signal $\Phi^{(0)}(z=L,t)$, where $\Phi^{(0)}(z=L,t)$ is the input signal $X(t)$ evolved at $z=L$ by the NLSE (1) with zero noise and $\beta=0$:

$$x_0(t_j)+iy_0(t_j)=\left[Y(t_j)-\Phi^{(0)}(L,t_j)\right]\exp\{-i\mu_j-i\phi_{0,j}\}=Y(t_j)\exp\{-i\mu_j-i\phi_{0,j}\}-\rho_j,\tag{12}$$

where we have used that for the zero noise and $\beta=0$ the NLSE (1) with the input condition $\Phi^{(0)}(z=0,t_j)=X(t_j)=\rho_j\exp\{i\phi_{0,j}\}$ has the solution $\Phi^{(0)}(z,t_j)=\rho_j\exp\{i\phi_{0,j}+i\mu_j z/L\}$. For the PDF (11) the factor Λ denoted as $\Lambda^{(0)}$ for the nondispersive channel reads

$$\Lambda^{(0)}=\prod_{j=0}^{M'-1}\frac{\delta_t}{\pi QL\sqrt{1+\mu_j^2/3}}.\tag{13}$$

To present the first correction to $P[Y|X]$ in small dispersion parameter $\widetilde{\beta}$ we first introduce the real functions $x(t)$ and $y(t)$ from analogy with the definition (12):

$$x(t_j) + iy(t_j) = [Y(t_j) - \Phi(z = L, t_j)]\, e^{-i\mu_j - i\phi_{0,j}}, \tag{14}$$

where $\Phi(z, t_j)$ is the solution of the NLSE equation (1) with zero noise, with nonzero β, and with the input boundary condition $\Phi(z = 0, t_j) = X(t_j) = \rho_j \exp\{i\phi_{0,j}\}$. The solution $\Phi(z, t)$ has the form: $\Phi(z,t) = e^{i\mu(t)z/L + i\phi_0(t)} \sum_{n=0}^{\infty} \widetilde{\Phi}^{(n)}(z,t)$. Quantities $\widetilde{\Phi}^{(n)}(z,t)$ up to the factors $(z\beta)^n$ are polynomials in nonlinear parameter $\mu(t)\frac{z}{L}$. The function $\widetilde{\Phi}^{(0)}(z,t) = \rho(t)$ corresponds to the leading order in small $\widetilde{\beta}$, and $\widetilde{\Phi}^{(1)}(z,t)$ corresponds to the next-to-leading order, and so on. As an example, the first correction reads $\widetilde{\Phi}^{(1)}(z,t) = \beta z \sum_{k=0}^{2} \phi_k^{(1)}(t) \left(\mu(t)\frac{z}{L}\right)^k$, where $\phi_0^{(1)}(t) = 2\dot{\rho}\,\dot{\phi}_0 + \ddot{\phi}_0\,\rho + i\left(\rho\dot{\phi}_0^{\,2} - \ddot{\rho}\right)$, $\phi_1^{(1)}(t) = \ddot{\rho} + 3\dot{\rho}^2/\rho + i\left(\ddot{\phi}_0\,\rho + 4\dot{\rho}\dot{\phi}_0\right)$, $\phi_2^{(1)}(t) = i\frac{2}{3}\left[\ddot{\rho} + 5\dot{\rho}^2/\rho\right]$. In the perturbation expansion it is easy to find $\widetilde{\Phi}^{(n)}(z,t)$ for any given n.

It is easy to find perturbatively $\Psi_{cl}(z,t)$ as well, and the corrections in $\widetilde{\beta}$ for the action $S[\Psi_{cl}]$ are presented as $S[\Psi_{cl}] = S^{(0)} + \delta S = S^{(0)} + S^{(1)} + S^{(2)} + \mathcal{O}(\widetilde{\beta}^3)$, here

$$S^{(0)} = \delta_t \sum_{j=0}^{M'-1} \frac{(1 + 4\mu_j^2/3)x_j^2 - 2\mu_j x_j y_j + y_j^2}{L(1 + \mu_j^2/3)}, \tag{15}$$

where $x_j \equiv x(t_j)$ and $y_j \equiv y(t_j)$ are given by (14). The first correction $S^{(1)}$ reads simpler in the continuous notations:

$$\begin{aligned} S^{(1)} = \int_T dt \frac{\beta}{15\rho^2(3+\mu^2)^4} \Big\{ & 2\mu\dot{\rho}^2\Big[(154\mu^8 + 1848\mu^6 + 5805\mu^4 + 9990\mu^2 - 405)x^2 + \\ & \mu(200\mu^8 + 1947\mu^6 + 5841\mu^4 + 7101\mu^2 - 5265)xy - 3(50\mu^8 + 492\mu^6 + 1593\mu^4 + \\ & 2322\mu^2 + 135)y^2\Big] + 2\mu(\mu^2+3)\rho\Big[\ddot{\rho}(3(10\mu^6 + 66\mu^4 + 135\mu^2 - 135)x^2 + \mu(40\mu^6 + 243\mu^4 + \\ & 288\mu^2 - 675)xy - 3(10\mu^6 + 66\mu^4 + 117\mu^2 - 45)y^2) + 3\dot{\rho}(10(\mu^2+3)\dot{\phi}_0((6\mu^3 + 33\mu)x^2 + \\ & 4(2\mu^4 + 7\mu^2 - 3)xy - 3\mu(2\mu^2+7)y^2) + \dot{y}(\mu(13\mu^4 + 102\mu^2 + 45)x - 6(3\mu^4 + 22\mu^2 + 15)y)) + \\ & \dot{\rho}\dot{x}(3\mu(11\mu^2(\mu^2+6) - 45)y - 2(14\mu^6 + 99\mu^4 + 135)x)\Big] + (\mu^2+3)^2\rho^2\Big[- 3\mu^4(x(-30x\ddot{\phi}_0 + \\ & 20\dot{x}\dot{\phi}_0 + 11\ddot{y}) - 40xy\dot{\phi}_0^{\,2} + y\ddot{x} + 30y^2\ddot{\phi}_0) + 6\mu^3(5x(\dot{\phi}_0(3x\dot{\phi}_0 + 2\dot{y}) + \ddot{x}) + y(80x\ddot{\phi}_0 + \\ & 10\dot{x}\dot{\phi}_0 + 3\ddot{y}) - 15y^2\dot{\phi}_0^{\,2}) - 45\mu^2(x(-11x\ddot{\phi}_0 + 6\dot{x}\dot{\phi}_0 + 4\ddot{y}) + 2y\dot{\phi}_0(\dot{y} - 4x\dot{\phi}_0) + 9y^2\ddot{\phi}_0) + \\ & 90\mu(y(-2x\ddot{\phi}_0 + 2\dot{x}\dot{\phi}_0 + \ddot{y}) + x\dot{\phi}_0(3x\dot{\phi}_0 + 2\dot{y}) - x\ddot{x} - 3y^2\dot{\phi}_0^{\,2}) - 135(-y(\ddot{x} - 2\dot{y}\dot{\phi}_0) + \\ & x(x\ddot{\phi}_0 + 2\dot{x}\dot{\phi}_0 + \ddot{y}) + y^2\ddot{\phi}_0) + 8\mu^5 x(\ddot{x} + 15y\ddot{\phi}_0)\Big]. \Big\} \end{aligned} \tag{16}$$

We present the normalization factor Λ in the form $\Lambda = \Lambda^{(0)} + \delta\Lambda = \Lambda^{(0)}\left\{1 + \frac{\Lambda^{(1)}}{\Lambda^{(0)}} + \frac{\Lambda^{(2)}}{\Lambda^{(0)}} + \mathcal{O}(\widetilde{\beta}^3)\right\}$. The normalization condition for conditional PDF $P[Y|X]$ reads

$$\int \mathcal{D}Y\, P[Y|X] = 1. \tag{17}$$

It is obvious that up to the next order in $\widetilde{\beta}$ one has

$$\frac{\Lambda^{(1)}}{\Lambda^{(0)}} = \left\langle \frac{S^{(1)}}{Q} \right\rangle_{P[Y|X]^{(0)}} \equiv \int \mathcal{D}Y\, P[Y|X]^{(0)} \frac{S^{(1)}}{Q}, \tag{18}$$

where $P[Y|X]^{(0)} = \Lambda^{(0)} \exp\left[-S^{(0)}/Q\right]$ with $S^{(0)}$ given by (15), and $\Lambda^{(0)}$ given by (13). Note that $\mathcal{D}Y = \prod_{j=0}^{M'-1} dReY(t_j)dImY(t_j) = \prod_{j=0}^{M'-1} dx_j dy_j$, see (14).

From the explicit discrete form of $S^{(1)}/Q$ one can immediately obtain $\Lambda^{(1)}/\Lambda^{(0)}$. Let us note that this quantity is singular in δ_t: in the continuous notations it has the form

$$\frac{\Lambda^{(1)}}{\Lambda^{(0)}} = \beta \int\limits_T dt \Big\{ -\delta_R''(0)\frac{2L\mu^3}{15\left(\mu^2+3\right)} - \delta_R(0)\frac{L\mu^2}{105\left(\mu^2+3\right)\rho^2}\Big\{14\left(2\mu^2+15\right)\rho\dot{\rho}\dot{\phi}_0 + \quad (19)$$
$$\mu\left(16\mu^2+189\right)\dot{\rho}^2 + 7\rho\left(\rho\left(2\mu\dot{\phi}_0^{\,2}+15\ddot{\phi}_0\right)+7\mu\ddot{\rho}\right)\Big\}.$$

The singularity is regularized by the renormalisation cutoff δt: the integral $\int dt$ is understood as the sum $\delta t \sum_{j=0}^{M'-1}$, $\int_T dt \delta_R(0)\ldots = \sum_{j=0}^{M'-1}\ldots$. The label "R" means regularization of delta-function in the fixed discretization scheme. In our scheme (10) the second derivative $\delta_R''(0)$ reads $\delta_R''(0) = -2/\delta t^3$. The small dispersive parameter of the perturbation expansion is $\widetilde{\beta} = -\beta L \delta_R''(0)/(2\delta_R(0))$. Another discretization scheme R' corresponds to another cutoff $\delta t_{R'} \neq \delta t$ and another relation between the noise bandwidth $W' \propto 1/\delta t_{R'}$.

In the second order in β we have

$$\frac{\Lambda^{(2)}}{\Lambda^{(0)}} = \left(\frac{\Lambda^{(1)}}{\Lambda^{(0)}}\right)^2 + \left\langle \frac{S^{(2)}}{Q} - \frac{1}{2}\left(\frac{S^{(1)}}{Q}\right)^2 \right\rangle_{P[Y|X]^{(0)}}. \quad (20)$$

We calculate this correction in the standard discretisation scheme. The result averaged over the angular variables is presented in the last section.

The conditional PDF $P[Y|X]$ has the following perturbative form in the first and the second orders in $\widetilde{\beta}$

$$P[Y|X] \approx P[Y|X]^{(0)}\left(1 + \left\{\frac{\Lambda^{(1)}}{\Lambda^{(0)}} - \frac{S^{(1)}}{Q}\right\} + \left\{\frac{\Lambda^{(2)}}{\Lambda^{(0)}} - \frac{\Lambda^{(1)}}{\Lambda^{(0)}}\frac{S^{(1)}}{Q} + \frac{1}{2}\left(\frac{S^{(1)}}{Q}\right)^2 - \frac{S^{(2)}}{Q}\right\}\right), \quad (21)$$

where the correction $\Lambda^{(1)}/\Lambda^{(0)}$ is given by (18), (19) and the correction $S^{(1)}$ to the action is given by (16).

The receiver (output signal detector) has bandwidth $W_r \ll W'$.

$$P_r[Y|X] = \int \left[\mathcal{D}Y\right]_{W'\backslash W_r} P[Y|X], \qquad \left[\mathcal{D}Y\right]_{W'\backslash W_r} = \prod_{k,\,\omega_k \in W'\backslash W_r} dReY_{\omega_k} dImY_{\omega_k}, \quad (22)$$

where PDF $P_r[Y|X]$ determines $P_{out}[Y]$, $H[Y]$, $H[Y|X]$, $I_{P[X]}$, C, etc. To estimate the mutual information $I_{P[X]}$ and the capacity $C = \max_{P[X]} I_{P[X]}$ we simplify our calculation of $P[Y|X]$ and $P_r[Y|X]$ by assumption $W' \sim W \sim W_r$. We will develop the perturbation expansion in $\widetilde{\beta} \ll 1$ on the base of $\beta = 0$ results. In real channels $W \ll W'$ but here we want only to demonstrate our method applied in the general case (of different W, W', W_r) as well.

3. Simple correlators

Let us consider the average of an arbitrary function $F(x(t), y(t'))$, see Eq. (14), over the realizations of the output signal $Y(t)$ when the input signal $X(t)$ is fixed:

$$\Big\langle F(x_j,y_k)\Big\rangle_{P[Y|X]} \equiv \int \mathcal{D}Y\, P[Y|X] F(x_j,y_k) = \Big\langle F(x_j,y_k)\Big\rangle_{(0)} + \Big\{\Big\langle \frac{S^{(1)}}{Q}\Big\rangle_{(0)} \Big\langle F(x_j,y_k)\Big\rangle_{(0)} - \Big\langle F(x_j,y_k)\frac{S^{(1)}}{Q}\Big\rangle_{(0)}\Big\} + \frac{\Lambda^{(1)}}{\Lambda^{(0)}}\Big\{\Big\langle \frac{S^{(1)}}{Q}\Big\rangle_{(0)} \Big\langle F(x_j,y_k)\Big\rangle_{(0)} - \Big\langle F(x_j,y_k)\frac{S^{(1)}}{Q}\Big\rangle_{(0)}\Big\} + \Big\{\Big\langle \frac{S^{(2)}}{Q}\Big\rangle_{(0)} \Big\langle F(x_j,y_k)\Big\rangle_{(0)} - \Big\langle F(x_j,y_k)\frac{S^{(2)}}{Q}\Big\rangle_{(0)}\Big\} - \frac{1}{2}\Big\{\Big\langle \Big(\frac{S^{(1)}}{Q}\Big)^2\Big\rangle_{(0)} \Big\langle F(x_j,y_k)\Big\rangle_{(0)} - \Big\langle F(x_j,y_k)\Big(\frac{S^{(1)}}{Q}\Big)^2\Big\rangle_{(0)}\Big\} + \mathcal{O}(\beta^3), \tag{23}$$

here we have used that the conditional PDF $P[Y|X]$ has the form (21). Here we introduced the average over "nonperturbed" conditional PDF: $\langle F(x_j,y_k)\Big\rangle_{(0)} = \int \mathcal{D}Y\, P[Y|X]^{(0)} F(x_j,y_k)$, where $P[Y|X]^{(0)} = \Lambda^{(0)} \exp\left[-S^{(0)}/Q\right]$, see (13) and (15). Note that $P[Y|X]^{(0)}$ does not correspond to the case $\beta = 0$, since $x = Re\left[(Y(t_j) - \Phi(L,t_j))\, e^{-i\mu_j - i\phi_{0,j}}\right] \neq x_0$, $y \neq y_0$ in (15).

For the illustration of our method we consider the following simple simultaneous correlators:

$$\alpha_{1,1}(t_j) = \Big\langle x(t_j)^2\Big\rangle_{P[Y|X]}, \qquad \alpha_{2,2}(t_j) = \Big\langle y(t_j)^2\Big\rangle_{P[Y|X]}, \qquad \alpha_{1,2}(t_j) = \Big\langle x(t_j)y(t_j)\Big\rangle_{P[Y|X]}. \tag{24}$$

Now we use (23), where all averages $\Big\langle \ldots \Big\rangle_{(0)}$ can be calculated using the Wick theorem [7] for pairing with Gaussian statistics of distribution $P[Y|X]^{(0)}$. The machinery calculation is simple and we present the final result in the first order in β:

$$\alpha_{1,1}(t_j) \approx \frac{QL}{2\,\delta_t}\Big\{1 - \frac{4\mu\beta L}{3\delta_t^2} - \frac{2\beta\mu L}{15\rho^2}\Big[10(\rho^2\dot{\phi}_0^{\,2} + \rho\ddot{\rho}) + \mu^2(10\rho\ddot{\rho} + 38\dot{\rho}^2) + 15\mu\rho^2\ddot{\phi}_0 + 45\mu\rho\dot{\rho}\dot{\phi}_0\Big]\Big\}, \tag{25}$$

$$\alpha_{2,2}(t_j) \approx \frac{QL(4\mu^2+3)}{6\,\delta_t}\Big\{1 - \frac{4\mu\,(4\mu^2-5)}{5\,(4\mu^2+3)\,\delta_t^2}\beta + \frac{2\beta L\mu}{105\rho^2(4\mu^2+3)}\Big[128\mu^4\dot{\rho}^2 + 196\mu^3\rho\dot{\rho}\dot{\phi}_0 + 84\mu^2\rho^2\dot{\phi}_0^{\,2} + 504\mu^2\rho\ddot{\rho} + 1932\mu^2\dot{\rho}^2 + 840\mu\rho^2\ddot{\phi}_0 + 1995\mu\rho\dot{\rho}\dot{\phi}_0 + 210\rho^2\dot{\phi}_0^{\,2} - 210\rho\ddot{\rho}\Big]\Big\}, \tag{26}$$

$$\alpha_{1,2}(t_j) \approx \frac{QL}{2\,\delta_t}\mu\Big\{1 - \frac{4\mu}{3\delta_t^2}\beta - \frac{\beta L}{45\rho^2}\Big[40\mu^3\rho\ddot{\rho} + 136\mu^3\dot{\rho}^2 + 60\mu^2\rho^2\ddot{\phi}_0 + 156\mu^2\rho\dot{\rho}\dot{\phi}_0 + 30\mu\rho^2\dot{\phi}_0^{\,2} - 75\mu\rho\ddot{\rho} - 135\mu\dot{\rho}^2 - 60\rho^2\ddot{\phi}_0 - 120\rho\dot{\rho}\dot{\phi}_0\Big]\Big\}, \tag{27}$$

where all functions in the r.h.s. of (25)–(27) are taken at $t = t_j$. Note that the correlators (25)–(27) are singular since they contain the noise power QL/δ_t. Here the dense grid spacing $\delta_t \propto 2\pi W'^{-1}$ is related with the bandwidth W' of the white noise. The dense grid spacing δ_t plays the role of the smallest ultraviolet cutoff of our field theory with the action (9). From this point of view all of quantities (25)–(27) require the renormalization procedure [7].

4. Mutual information and capacity

From representation of [5] the mutual information $I_{P[X]}$ in the leading order in 1/SNR reads:

$$I_{P[X]} = M \log\left[\frac{P_{ave}}{P_{noise}}\right] + \left\langle \log\left[\frac{\Lambda}{\Lambda_{QL}}\right]\right\rangle_{P[X]}, \tag{28}$$

where $\Lambda_{QL} = \left(\widetilde{\delta}_t/(\pi QL)\right)^M$, and $\left\langle F[X]\right\rangle_{P[X]} = \int \mathcal{D}X P[X]F[X]$ is averaging over the input signal PDF $P[X]$. The representation (28) was obtained in [5] on the assumption that PDF $P[Y|X]$ obeys the factorization property (for meaning and remnant channels, see [5]). In our estimations we assume that $W' \sim W \sim W_r$ (all time channels are meaning, $M' = M$) and (28) holds true in the leading order in 1/SNR (without any factorization assumption). For realistic case one has $W' \gg W$ and for the concrete output signal detector $W_r \neq W$. However, our estimations will give the general capacity behavior, and our method allows one to find the capacity for given realizations of noise (W') and output signal detector (W_r).

$$I_{P[X]} = M \log\left[\frac{P_{ave}}{P_{noise}}\right] - \frac{M}{2}\left\langle \int_T \frac{dt}{T} \log\left[1+\mu^2(t)/3\right]\right\rangle_{P[X]} + \left\langle \log\left[1+\frac{\delta\Lambda}{\Lambda^{(0)}}\right]\right\rangle_{P[X]}, \tag{29}$$

where the last term in (29) delivers the correction in β to the mutual information $I^{(\beta=0)}_{P[X]}$ (the first two terms) of an nondispersive channel [6]. In the first order in $\widetilde{\beta}$ we arrive at the representation

$$I_{P[X]} = I^{(\beta=0)}_{P[X]} + \left\langle \frac{\Lambda^{(1)}}{\Lambda^{(0)}}\right\rangle_{P[X]} + \left\langle \frac{\Lambda^{(2)}}{\Lambda^{(0)}} - \frac{1}{2}\left(\frac{\Lambda^{(1)}}{\Lambda^{(0)}}\right)^2\right\rangle_{P[X]} + \mathcal{O}(\widetilde{\beta}^3), \tag{30}$$

where $\Lambda^{(1)}/\Lambda^{(0)}$ is presented in (19) in the continuous notations. However for explicit averaging $\left\langle \ldots \right\rangle_{P[X]}$ it is more convenient to use the discrete notation for $\Lambda^{(1)}/\Lambda^{(0)}$ and $\Lambda^{(2)}/\Lambda^{(0)}$.

Now it is easy to see that the first correction in $\widetilde{\beta}$ in (30) vanishes for wide class of distributions $P[X]$. In particular, for an axially symmetric distribution $P[X] = P[|X|] = \prod_{j=0}^{M-1} P_j[\rho_j]$ the first correction in $\widetilde{\beta}$ is zero: $\left\langle \frac{\Lambda^{(1)}}{\Lambda^{(0)}}\right\rangle_{P[|X|]} = \int \prod_{j=0}^{M-1} \frac{\Lambda^{(1)}}{\Lambda^{(0)}} P_j[\rho_j]\rho_j d\rho_j d\phi_{0,j} = 0$.

For the channel capacity correction ΔC we need to know the second correction $\Lambda^{(2)}$ to the normalization factor, see Eq. (36) below. This correction $\Lambda^{(2)}$ has to be averaged over zero order optimal input signal distribution $P_X^{opt(\beta=0)}[X]$ known in the intermediate power region $QL \ll P \ll \left(Q\gamma^2 L^3\right)^{-1}$, see [6]. This distribution is angular symmetric, [6]:

$$P_X^{opt(\beta=0)}[X] = \prod_{j=0}^{M'-1} P_X^{opt(\beta=0),j}[X(t_j)] = \prod_{j=0}^{M'-1} N_0 \frac{\exp\left\{-\lambda_0 |X(t_j)|^2\right\}}{\sqrt{1+\mu^2(t_j)/3}}, \tag{31}$$

where N_0 and λ_0 are functions depending on average input signal power $P_{ave} = P/\delta t$ through the conditions, see Eqs. (46)–(47) in Ref. [6]:

$$2\pi N_0 \int_0^\infty \frac{d\rho\, \rho\, e^{-\lambda_0\rho^2}}{\sqrt{1+\gamma^2L^2\rho^4/3}} = 1, \qquad 2\pi N_0 \int_0^\infty \frac{d\rho\, \rho^3 e^{-\lambda_0\rho^2}}{\sqrt{1+\gamma^2L^2\rho^4/3}} = P/\delta t. \tag{32}$$

The typical average for our calculation of the average $\Lambda^{(2)}$ has the form of the integral

$$j^{opt(\beta=0)}(k,n) \equiv \int dReX_j dImX_j P_X^{opt(\beta=0),j}[X(t_j)] \frac{\mu^k(t_j)}{(3+\mu^2(t_j))} = n_0(\widetilde{\gamma}) 3^{k/2-n} \widetilde{j}(k,n;\alpha_0(\widetilde{\gamma})), \quad (33)$$

here k and n are integers, and $\widetilde{j}(k,n;\alpha) \equiv \int_0^\infty dz \frac{e^{-\alpha z} z^k}{(1+z^2)^{n+1/2}}$. The average $j^{(\beta=0)}(k,n)$ can be expressed through the functions $\alpha_0(\widetilde{\gamma}) \equiv P\lambda_0/(\delta t\widetilde{\gamma})$ and $n_0(\widetilde{\gamma}) \equiv \pi P N_0/(\delta t\widetilde{\gamma})$ which depends on dimensionless parameter $\widetilde{\gamma} = \gamma LP/(\sqrt{3}\delta t)$ only. The properties of these functions were studied in details in Ref. [6]. For integer k and n and arbitrary α the integral $\widetilde{j}(k,n;\alpha)$ can be recurrently expressed through $\widetilde{j}(0,0;\alpha) = \int_0^\infty dz\, e^{-\alpha z}/\sqrt{1+z^2} = \frac{\pi}{2}\Big(H_0(\alpha) - Y_0(\alpha)\Big)$ with $Y_0(\alpha)$ and $H_0(\alpha)$ being the Neumann and Struve functions of zero order, respectively. We will use this function $\widetilde{j}(k,n;\alpha)$ for the calculation of the channel capacity corrections. The explicit form of the correction $\Lambda^{(2)}$ is very cumbersome to be presented here, and we present the discrete form of the correction $\Lambda^{(2)}$ averaged over the angular variables in the following form which is the most convenient for further numerical calculation of ΔC:

$$\left\langle \frac{\Lambda^{(2)}}{\Lambda^{(0)}} \right\rangle_{P[|X|]} = 4M'\widetilde{\beta}^2 Re\Bigg[\frac{1}{M'}\sum_{j=0}^{M'-1}\Bigg\{6J^{j+1,j}_{(0,2,0,1,4)} - 12J^{j+1,j}_{(0,2,0,1,5)} + 18J^{j+1,j}_{(0,2,1,1,6)} - 4iJ^{j+1,j}_{(0,3,0,1,4)} +$$
$$24iJ^{j+1,j}_{(0,3,0,1,5)} - 48iJ^{j+1,j}_{(0,3,0,1,6)} - iJ^{j+1,j}_{(0,3,1,1,5)} + 54iJ^{j+1,j}_{(0,3,1,1,6)} - 72iJ^{j+1,j}_{(0,3,1,1,7)} - 36J^{j+1,j}_{(1,1,0,2,4)} +$$
$$72J^{j+1,j}_{(1,1,0,2,5)} + 72J^{j+1,j}_{(1,1,0,2,6)} + 72iJ^{j+1,j}_{(1,2,0,2,4)} - 432iJ^{j+1,j}_{(1,2,0,2,5)} + 648iJ^{j+1,j}_{(1,2,0,2,6)} + 432iJ^{j+1,j}_{(1,2,0,2,7)} +$$
$$18iJ^{j+1,j}_{(1,2,1,1,6)} - 108iJ^{j+1,j}_{(1,2,1,1,7)} + 16J^{j+1,j}_{(1,3,0,1,5)} - 72J^{j+1,j}_{(1,3,0,1,6)} + 96J^{j+1,j}_{(1,3,0,1,7)} + 30J^{j+1,j}_{(1,3,0,2,4)} -$$
$$300J^{j+1,j}_{(1,3,0,2,5)} + 1068J^{j+1,j}_{(1,3,0,2,6)} - 1008J^{j+1,j}_{(1,3,0,2,7)} - 1584J^{j+1,j}_{(1,3,0,2,8)} + 6J^{j+1,j}_{(1,3,1,1,5)} - 84J^{j+1,j}_{(1,3,1,1,6)} +$$
$$396J^{j+1,j}_{(1,3,1,1,7)} - 504J^{j+1,j}_{(1,3,1,1,8)} - 16iJ^{j+1,j}_{(1,4,0,1,5)} + 176iJ^{j+1,j}_{(1,4,0,1,6)} - 768iJ^{j+1,j}_{(1,4,0,1,7)} +$$
$$1344iJ^{j+1,j}_{(1,4,0,1,8)} + 12iJ^{j+1,j}_{(1,4,0,2,4)} - 24iJ^{j+1,j}_{(1,4,0,2,5)} - 312iJ^{j+1,j}_{(1,4,0,2,6)} + 1296iJ^{j+1,j}_{(1,4,0,2,7)} -$$
$$576iJ^{j+1,j}_{(1,4,0,2,8)} - 3456iJ^{j+1,j}_{(1,4,0,2,9)} + 32J^{j+1,j}_{(1,5,0,1,6)} - 320J^{j+1,j}_{(1,5,0,1,7)} + 1344J^{j+1,j}_{(1,5,0,1,8)} -$$
$$2304J^{j+1,j}_{(1,5,0,1,9)} + 8J^{j+1,j}_{(1,5,0,2,4)} - 96J^{j+1,j}_{(1,5,0,2,5)} + 336J^{j+1,j}_{(1,5,0,2,6)} - 96J^{j+1,j}_{(1,5,0,2,7)} - 1536J^{j+1,j}_{(1,5,0,2,8)} +$$
$$6912J^{j+1,j}_{(1,5,0,2,10)} + 16iJ^{j+1,j}_{(1,6,0,2,5)} - 176iJ^{j+1,j}_{(1,6,0,2,6)} + 768iJ^{j+1,j}_{(1,6,0,2,7)} - 1344iJ^{j+1,j}_{(1,6,0,2,8)} -$$
$$32J^{j+1,j}_{(1,7,0,2,6)} + 320J^{j+1,j}_{(1,7,0,2,7)} - 1344J^{j+1,j}_{(1,7,0,2,8)} + 2304J^{j+1,j}_{(1,7,0,2,9)} + 18J^{j+1,j}_{(2,0,1,1,6)} -$$
$$18iJ^{j+1,j}_{(2,1,1,1,6)} + 108iJ^{j+1,j}_{(2,1,1,1,7)} - 6J^{j+1,j}_{(2,2,1,1,4)} + 12J^{j+1,j}_{(2,2,1,1,5)} + 24J^{j+1,j}_{(2,2,1,1,6)} + 72J^{j+1,j}_{(2,2,1,1,7)} -$$
$$432J^{j+1,j}_{(2,2,1,1,8)} + 2iJ^{j+1,j}_{(2,3,1,1,4)} - 10iJ^{j+1,j}_{(2,3,1,1,5)} - 48iJ^{j+1,j}_{(2,3,1,1,6)} + 588iJ^{j+1,j}_{(2,3,1,1,7)} -$$
$$1728iJ^{j+1,j}_{(2,3,1,1,8)} + 1728iJ^{j+1,j}_{(2,3,1,1,9)} - 6iJ^{j+1,j}_{(3,0,1,1,5)} + 18iJ^{j+1,j}_{(3,0,1,1,6)} - 6J^{j+1,j}_{(3,1,1,1,5)} +$$
$$48J^{j+1,j}_{(3,1,1,1,6)} - 108J^{j+1,j}_{(3,1,1,1,7)} - 2iJ^{j+1,j}_{(3,2,1,1,4)} + 14iJ^{j+1,j}_{(3,2,1,1,5)} - 36iJ^{j+1,j}_{(3,2,1,1,6)} + 12iJ^{j+1,j}_{(3,2,1,1,7)} +$$
$$20J^{j+1,j}_{(3,3,1,1,6)} - 288J^{j+1,j}_{(3,3,1,1,7)} + 1752J^{j+1,j}_{(3,3,1,1,8)} - 4608J^{j+1,j}_{(3,3,1,1,9)} + 4608J^{j+1,j}_{(3,3,1,1,10)} -$$
$$12J^{j+1,j}_{(4,0,1,1,6)} + 72J^{j+1,j}_{(4,0,1,1,7)} - 72J^{j+1,j}_{(4,0,1,1,8)} + 12iJ^{j+1,j}_{(4,1,1,1,6)} - 144iJ^{j+1,j}_{(4,1,1,1,7)} +$$
$$576iJ^{j+1,j}_{(4,1,1,1,8)} - 576iJ^{j+1,j}_{(4,1,1,1,9)} + 4J^{j+1,j}_{(4,2,1,1,6)} - 96J^{j+1,j}_{(4,2,1,1,7)} + 744J^{j+1,j}_{(4,2,1,1,8)} -$$
$$2304J^{j+1,j}_{(4,2,1,1,9)} + 2304J^{j+1,j}_{(4,2,1,1,10)}\Bigg\}\Bigg], \quad (34)$$

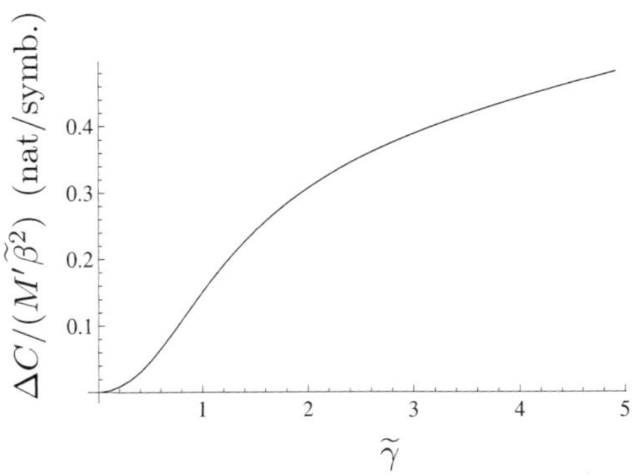

Figure 1. The correction ΔC to the channel capacity as a function of parameter $\widetilde{\gamma} = \frac{\gamma LP}{\delta_t\sqrt{3}}$ for low powers;

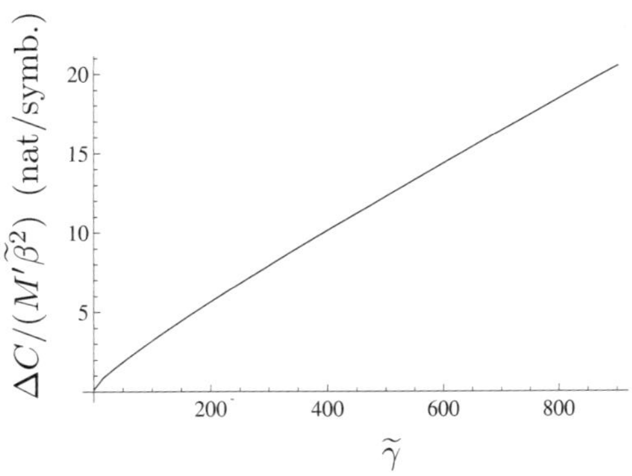

Figure 2. The correction ΔC to the channel capacity as a function of parameter $\widetilde{\gamma} = \frac{\gamma LP}{\delta_t\sqrt{3}}$ for high powers.

where we have used the notations

$$J^{j+1,j}_{(k_1,k_2,n_1,n_2,k)} = \left\langle \frac{\mu^{k1}(t_{j+1})\mu^{k2}(t_j)}{[3+\mu^2(t_{j+1})]^{n_1}[3+\mu^2(t_j)]^{n_2}} \mathrm{expc}_k\left(i\mu(t_{j+1}) - i\mu(t_j)\right) \right\rangle_{P[|X|]}, \tag{35}$$

here $\mathrm{expc}_k(z) = \int_0^1 dz' e^{zz'}(1-z')^{k-1}/(k-1)!$ is truncated exponent series.

To obtain the first non-vanishing correction ΔC to the capacity one should calculate the normalization factor Λ with the accuracy β^2:

$$\Delta C = \int \mathcal{D}X\, P_X^{opt(\beta=0)}[X]\, \frac{\Lambda^{(2)}}{\Lambda^{(0)}}. \tag{36}$$

To find ΔC we should calculate averages (35) for the zero order optimal input signal distribution $P_X^{opt(\beta=0)}[X]$, see Eq. (31). It is easy to express averages (35) through the known functions $n_0(\widetilde{\gamma})$, $\alpha_0(\widetilde{\gamma})$ and integral $\widetilde{j}(k,n;\alpha)$:

$$J^{j+1,j}_{(k_1,k_2,n_1,n_2,k)}\Big|_{P[X]=P_X^{opt(\beta=0)}[X]} = 3^{k_1/2+k_2/2-n_1-n_2} n_0^2(\widetilde{\gamma}) \int_0^1 dz' \frac{(1-z')^{k-1}}{(k-1)!} \times$$
$$\widetilde{j}(k_1,n_1;\alpha_0(\widetilde{\gamma}) - i\sqrt{3}z')\widetilde{j}(k_2,n_2;\alpha_0(\widetilde{\gamma}) + i\sqrt{3}z'), \tag{37}$$

where for given dimensionless parameter $\widetilde{\gamma}$ the average $J^{j+1,j}_{(k_1,k_2,n_1,n_2,k)}\Big|_{P[X]=P_X^{opt(\beta=0)}[X]}$ does not depend on the time t_j and the integral (37) can be easily calculated numerically. This means that the formula (34) for $J^{j+1,j}_{(k_1,k_2,n_1,n_2,k)}$ from Eq. (37) determines the correction ΔC as the function of parameter $\widetilde{\gamma}$ and we can omit $\frac{1}{M'}\sum_0^{M'-1}$ therein. Note that representations (34) is not unique.

Repeating the variation procedure introduced in [6] it is easy to see that taking into account the corrections proportional to β in Λ one obtains that the optimal input signal distribution $P_X^{opt}[X]$ acquires an angular dependance, but the correction to the capacity proportional to β is still zero.

The numerical results of our estimations of the capacity correction ΔC are shown in Fig. 1 and Fig. 2 for low powers ($\widetilde{\gamma}_{noise} < \widetilde{\gamma} < 5$) and high powers ($1 < \widetilde{\gamma} < 800$), correspondingly, both calculated within the intermediate power regime: $\widetilde{\gamma}_{noise} \ll \widetilde{\gamma} \ll (\widetilde{\gamma}_{noise})^{-1}$. Here $\widetilde{\gamma}$ is the dimensionless nonlinearity parameter $\widetilde{\gamma} = \gamma LP/(\delta_t\sqrt{3})$, $\widetilde{\gamma}_{noise} = \gamma L^2 Q/(\delta_t\sqrt{3})$, so that

$\mathrm{SNR} = \widetilde{\gamma}/\widetilde{\gamma}_{noise}$. In Fig. 1 and Fig. 2 we present the ratio $\Delta C/(M'\widetilde{\beta}^2)$, where $\widetilde{\beta} = \beta L/\widetilde{\delta}_t^2$, and M' is total time mesh points (independent time channels). The correction ΔC is positive in the intermediate power regime, i.e., it enhances the channel capacity of a nondispersive channel [6]. For typical fiber optical links [8] one has: $L = 1000\,\mathrm{km}$, $\gamma = 1.31(\mathrm{Wkm})^{-1}$, $2\pi/\delta_t = 100\,\mathrm{GHz}$, $P_{noise} = QL/\delta_t = 5.3 \times 10^{-4}\mathrm{mW}$. For these parameters one has $\widetilde{\gamma}_{noise} \approx 4 \times 10^{-7}$, and our results are reliable if $\widetilde{\beta} \ll 1$, i.e., $\beta \ll 1\,\mathrm{ps}^2/\mathrm{km}$.

5. Concluding remarks

We present the path-integral approach [3, 5, 6, 9] to the calculation of the information theory quantities (conditional probability density function $P[Y|X]$, output signal correlators, mutual information $I_{P[X]}$, optimal input signal distribution $P_X^{opt}[X]$, channel capacity C, and so on) for the communication channel modelled by the nonlinear Shrödinger equation with additive white Gaussian noise at large signal-to-noise ratio. We apply this approach to the channel with the small dispersion parameter and find the first and the second corrections to $P[Y|X]$ perturbatively. Then we apply our method to the calculation of the simple output signal correlators. Finally on the simple model assumptions (the noise bandwidth W' is of the same order as the input signal bandwidth W) we consider the estimations for the channel capacity correction ΔC in small dispersion parameter in the intermediate power regime. Of course, this model assumptions are far from the realistic channel but our approach allows one the calculate the correlators, the mutual information and the channel capacity corrections for small dispersion in general case (for various W', W and various output signal detectors with the bandwidth W_r) as well.

Acknowledgments

Acknowledgments Part of the work, i.e., the calculation of the conditional PDF $P[Y|X]$ and the channel capacity estimations for small dispersion, was supported by the Russian Foundation for Basic Research (RFBR), Grant No. 16-31-60031. The calculations of the small dispersion corrections to output signal correlators have been performed with support of the the Russian Science Foundation (RSF), Grant No. 16-11-10133.

References

[1] Shannon C E 1949 *Communication in the presence of noise. Proc. Institute of Radio Engineers* vol 37 p 1
[2] Martin P C, Siggia E D and Rose H A 1973 *Phys. Rev.* A **8** 423
[3] Terekhov I S, Vergeles S S and Turitsyn S K 2014 *Phys. Rev. Lett.* **113** 230602
[4] Hochberg D, Molina-Paris C, Perez-Mercader J and Visser M *Phys. Rev.* E **60** 6343
[5] Terekhov I S, Reznichenko A V and Turitsyn S K 2016 *Phys. Rev.* E **94** 042203
[6] Terekhov I S , Reznichenko A V, Kharkov Ya A and Turitsyn S K 2017 *Phys. Rev.* E **95** 062133
[7] Itzykson C, Zuber J B 1980 *Quantum Field Theory* (New York: McGraw-Hill Book Company)
[8] Essiambre R-J, Kramer G, Winzer P J, Foschini G J and Goebel B 2010 *J. of Lightwave Technol.* **28** 662
[9] Panarin A A, Reznichenko A V and Terekhov I S 2016 *Phys. Rev.* E **95** 012127

Smart Photonic Carbon Brush: FBG Length as Sensing Parameter

O.G. Morozov[1], I.I. Nureev[1], A.A. Kuznetsov[1]and V.I. Artemiev[1]
[1]Radiophotonics and Microwave Technologies, Kazan National Research Technical University n.a. A.N.Tupolev-KAI, 10, K. Marx St., 420111, Kazan, Russia. Contact Phone: +79172663227

E-mail: microoil@mail.ru

Abstract. This article deals with problem of carbon brush`s length measurements. There are many applications where regular inspection is not feasible because of a number of factors including, for example, time, labor, cost and disruptions due to down time. Thus, there is a need for a system that can monitor the brush`s length to calculate it`s wear rate, while the component is in operation or without removing of the component from its operational position. We propose a novel method for characterization of carbon brush`s length. This method based on the usage of advantages of the multiplicative response of FBGs and FBG arrays: spectral parameters depend on several aspects, such as grating`s period, refractive index, it`s physical length and so on. We are the first, in our point of view, who proposed to use third parameter for sensing application and prospectively all three parameters for complex measurement: the change of FBG`s length is used to measure length of the brush and it`s wear rate, grating`s central wavelength shift for temperature (due to refractive index change) and mechanical stress (due to grating`s period variations) measurements. The results of modelling and experiments are presented.

1. Introduction

Surface-reducing wear occurs in machines by frictional contact of adjacent parts and by other forms of erosion. For example, turbine engines have bearings, shroud liners, combustor liner spring clips, and other areas where reduction wear occurs. Detecting such wear can be critical to safe operation of the machine [1]. The clearest example of the surface-reducing wear in electric machines is carbon brush. Modern brushes are mentioned for high speed and high current applications and have unique ability to share current at minimal frictions in order to provide cooler running and outstanding brush life. Despite of that, all manufacturers offer carbon brush wear detection systems, making maintenance operation easier.

Sensors have been designed to detect wear without disassembly of components. For example, electrical conductors may be embedded in a wear surface. A detector connected to these conductors senses an open circuit caused by wear through a conductor, and determines a wear depth [2]. Alternate device for wear magnitude measurement in friction is organized, for example, by positioning in product two control electric conductors, isolated at a certain depth from the product body [3]. Here the threshold wear value of the product determined by the depth location of control electrical conductors is controlled. When the determined depth reached friction products break down the isolation of the conductors which short-circuit by the elements of the second friction products and the registration system receives an electrical signal of reaching the wear threshold. A disadvantage of such devices is the possibility of high voltage entering at control electrical conductors when the friction product is under it.

To eliminate this drawback, the control conductors are made by fibre optic. Patent [4] describes loops of optical fibres embedded in a wear surface. Each loop forms an optical circuit with a light source at one end and a detector at the other end. Reduction wear breaks one or more loops in succession, which is detected by respective losses of signal. This design requires enough sensor area for the embedded fibres to form loops, and requires both a source leg and a detector leg of each fibre loop to pass through the sensor block to the source and detector. Additional disadvantages of the device and the above mentioned one are the lack of temperature control and provision only the threshold wear index indication, but not its rate.

Our analysis showed that in practice there are exist only separate methods of wear and friction surfaces temperature measurement. For joint temperature and wear measurement was found a technology1, the point of which is that of the control fibre optic light guide is placed in the brush at a depth equal to or less than the distance to the friction surface, at which end the section from phosphor is formed. Phosphor is excited by a pulsed laser source probe and emits light energy at different wavelength because of luminescence for example. To measure the wear and temperature product value at friction the phosphor characteristics are used. The detector is characterized by the possibility of amplitude and spectral measurements with incoming probe wavelengths determination, determines the wear value based on the amplitude characteristics of re-emission, changing, when phosphor is being erased, and temperature based on the spectral characteristics of re-emission, namely, temperature-dependent wavelength re-emission. Note the obvious complexity of this joint technology and the possibility of its implementation on the narrow thickness of the phosphor, which leads to the necessity of using control fibre optic light guides at different depths.

Patent [5] also deals with an optical fibre in a wear surface. Light injected into the fibre reflects off the opposed abrading surface and returns through the same fibre. The fibre tip wears along with the wear surface, and the light circuit length is reduced. This length is measured by interferometry to determine a wear depth. The problems of this invention are necessity to analyse reflection from the abrading surface, which requires signal analysis that deals with variable reflectivity, and application of interferometry, thus eliminating elements such as reference beam optics. For temperature measurements, authors use additional fibre Bragg grating (FBG) sensor embedded in brush parallel to main fibre.

In present paper, we demonstrate smart photonic FBG sensor for brush wear monitoring with additional functions of brush wear rate, temperature and hanging monitoring with possibility to rotor speed measuring. We are first from our point of view who shown how to measure brush wear by means of FBG length monitoring and use of other FBG multiplicative properties to measure above mentioned characteristics. In second part, we analyse FBG characteristics. In third part we present modelling results and in fourth its practical confirmation. In conclusion, we will resume results of above mentioned rotor speed measurements and discuss the directions of researches development.

2. FBG`s length as sensing parameter. New way for fibre optic sensors

In our previous article [6] we proposed a new sensing method, based on dependence of FBG spectral characteristics on its length.

The FBG spectra in respect to reflection R is expressed through its length L and detuning δ [7]:

IOP Conf. Series: Journal of Physics: Conf. Series **999** (2018) 012017 doi:10.1088/1742-6596/999/1/012017

$$R = \frac{\sinh^2\left[\kappa L\sqrt{1-(\delta/\kappa)^2}\right]}{\cosh^2\left[\kappa L\sqrt{1-(\delta/\kappa)^2}\right]-(\delta/\kappa)^2}, \tag{1}$$

where κ – the coupling coefficient of the forward and reverse modes, (δ/κ) – relative detuning. FBG detuning with period Λ equals to $\delta = \Omega - (\pi/\Lambda)$, where $\Omega = 2\pi n_{eff}/\lambda$.

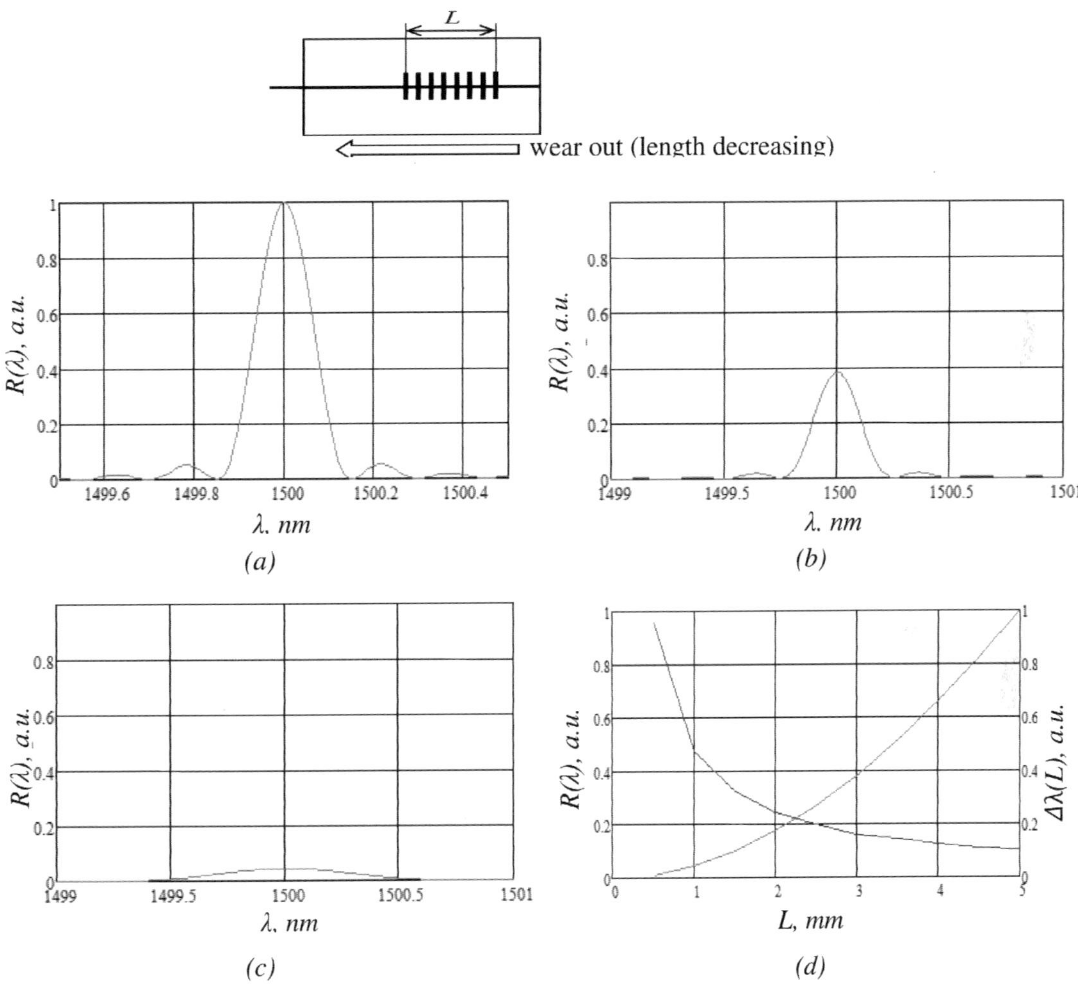

Figure 1. FBG spectral characteristics:
a – reflectance for FBG of 5 mm length; *b* – reflectance for FBG of 3 mm length; *c* – reflectance for FBG of 1 mm length; *d* – reflectance *R* (red line) and FWHM *Δλ* (blue line) as function of FBG length *L*.

To illustrate the changings in FBG spectral properties we present the dependence of the reflectance on the wavelength for several FBG lengths as shown on Figure 1, (a)-(c). The effective refractive index of the fundamental mode was chosen n_{eff} = 1.5. The grating period was selected so that the core for the unperturbed resonant wavelength matched λ_{FBG}=1500 nm. The initial length of the grating was chosen to be $L = 5$ mm, $\kappa \cdot L = 0.38$. In the numerical modeling of the FBG spectral properties, we reduced the length of the grating. As the result, we show on Figure 1 (d) the dependence of the reflectance (R) and the spectral width ($\Delta\lambda$) from the length (L) of the grating.

Typical length of FBG is 3-12 mm. In some cases, when the length of the measured product can reach tens of millimeters, this range of measurement becomes insufficient. In these cases we can offer other different structures, based on FBGs, such as phase shifted FBGs and Fabri-Perot interferometer, based on two FBGs [8]. For the first structure, measurement range can be increased to ~24 mm, for the second – up to ~36 mm.

Typical length of carbon brushes begins from several millimeters (for electric tools) to 60-70 mm for motor-wheels of dumpers and trains. It`s evident, that earlier structures can`t provide such measurement range, so here we propose to use FBG array, the structure, that consists of many identical FBGs fabricated in same fiber with a specific step. In this case, we have a number of trickles (3-12 mm length), within which we can ensure continuous measurement with free spaces between them. The use of an array of the same type FBGs allows the maximum effective use of the operational spectral range of the interrogator. The number of sensors located in the body of the measuring object is determined, firstly, by its length, and, secondly, by the requirement for the number of controllable wear levels (the areas where it is required to monitor the wear value with high accuracy).We called this quasi continuous length measurements.

3. FBG array for quasi-continuous length measurements

FBG arrays were described mathematically in [9]. The change in the spectrum response of the FBG array is caused by various types of crosstalk [9]. The first one is the spectral shadowing $I_{ri}(\lambda)$, which describes the phenomenon of the distortion of the reflection spectrum of the FBG (see Figure 2) at the near end (FBG_1 and FBG_2), caused by the insertion loss of reflected radiation from the FBG at the far end (FBG_{N-1} and FBG_N). The second one is cross-distortions $C_{ri}(\lambda)$ caused by the multiplicity of the response associated with a large number of multiple reflection crosstalk. The resulting reflection spectrum from the i-th FBG is the sum of $I_{ri}(\lambda)$ and $C_{ri}(\lambda)$ (2).

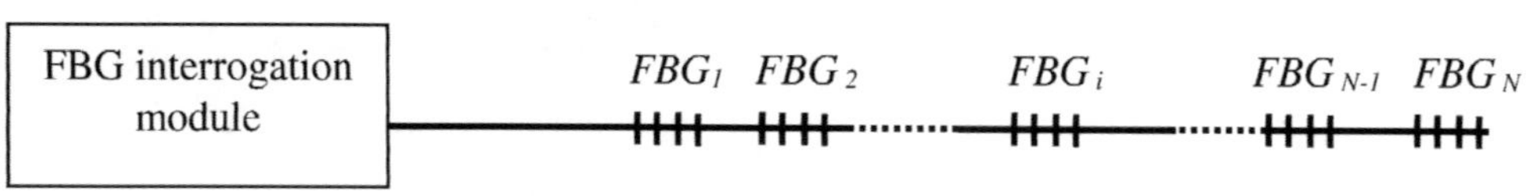

Figure 2. FBG array.

$$R_{AR}(\lambda, N) = I_{ri}(\lambda) + C_{ri}(\lambda), \tag{2}$$

where N – total number of FBGs; $I_{ri}(\lambda) = [1 - R\ \ (\lambda)]^{2(N-1)} R\ \ (\lambda)$; $C_{ri}(\lambda) = \frac{(N-1)(N-2)}{2} R\ \ (\lambda)^3 [1 - R\ \ (\lambda)]^{2N-4}$, $N \geq 3$, $R\ \ (\lambda)$ described as (1).

The resulting reflection spectrum of the FBG array is the sum of the radiation reflected from the N FBGs (3):

$$R_{AR}(\lambda, N) = \sum_{i=1}^{N} R_{ARi}(\lambda, N) = \sum_{i=1}^{N} [I_{ri}(\lambda) + C_{ri}(\lambda)]. \tag{3}$$

For the Mathcad simulation, FBGs with the Gaussian reflection profiles, described in higher section, were selected. The number of gratings $N=4$ was chosen. The initial profile of the reflection spectrum of the array is shown in Figure 3 (a).

In the course of modeling the wearing of the FBG array, the resultant reflection spectrum was constructed according to the expression (3) for a different number of gratings: N = 4 (initial spectrum), 3 and 1, which are shown on Figure 3 (a)-(c). The change in the FBG spectrum at those times when any sensor wearing occurs is described in previous section.

According to these results, we can conclude that as the number of FBGs in the array decreases, the reflection coefficient $R_{AR}(\lambda)$ decreases and the spectral width $\Delta\lambda_{AR}(\lambda)$ slightly changes, as shown on Figure 3 (d).

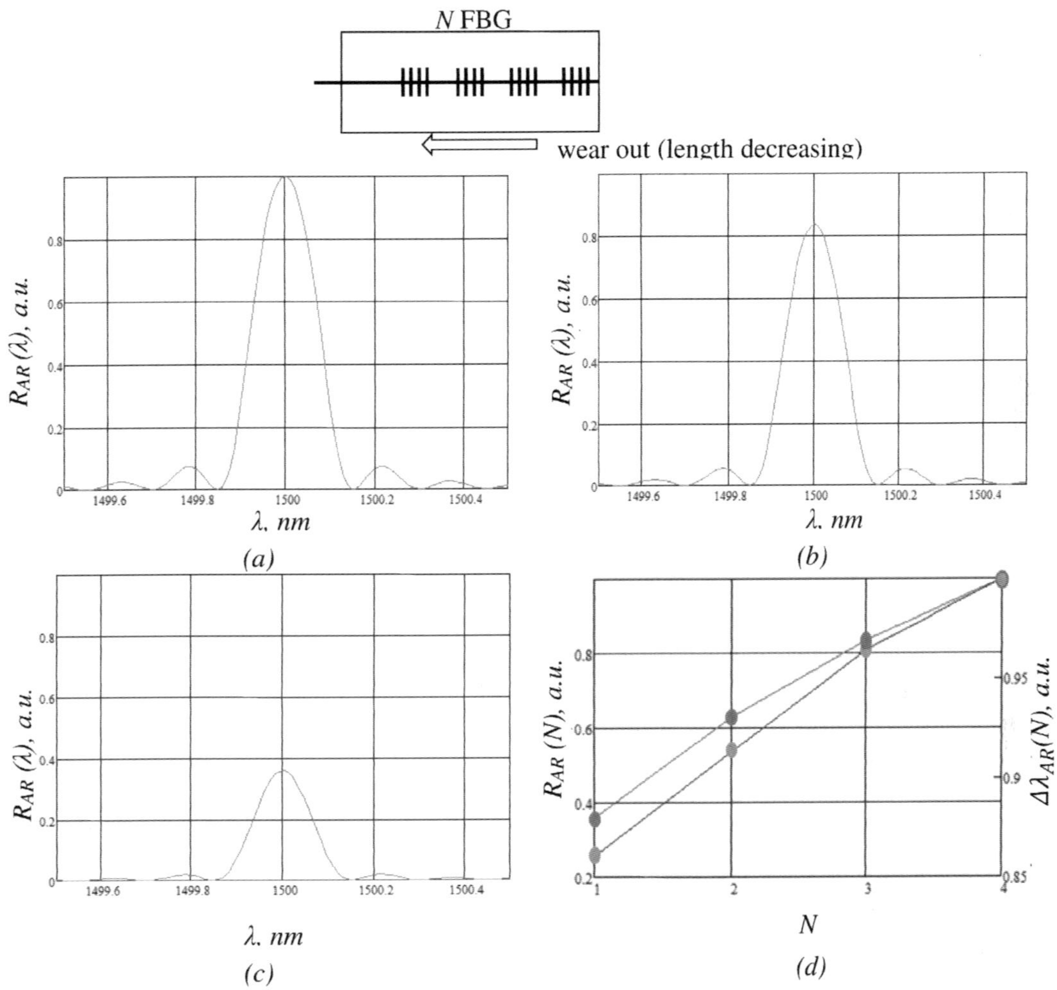

Figure 3. FBG array spectral characteristics:
a – reflectance for *N=4* FBG array; *b* – reflectance for *N=3* FBG array; *c* – reflectance for *N=1* FBG array; *d* – reflectance R_{AR} (red line) and FWHM $\Delta\lambda_{AR}$ (blue line) as function of total number of FBG *N*.

4. FBG wear sensor interrogation technique

There are many ways FBG interrogation. All of these methods can be divided into two classes: with a direct definition of the spectrum parameters [10-11] and radiophotonic [12-13].

These methods allow high-precision measurements, but these devices are very expensive. The requirements for the measurements accuracy (±2,5% of total length, according to state standards) is much lower than those achieved by these methods, therefore, an alternative scheme of the interrogation device implementing the power comparison method and consisting of the measuring and two reference FBGs can be proposed.

Structural scheme of such devise is shown on Figure 4.

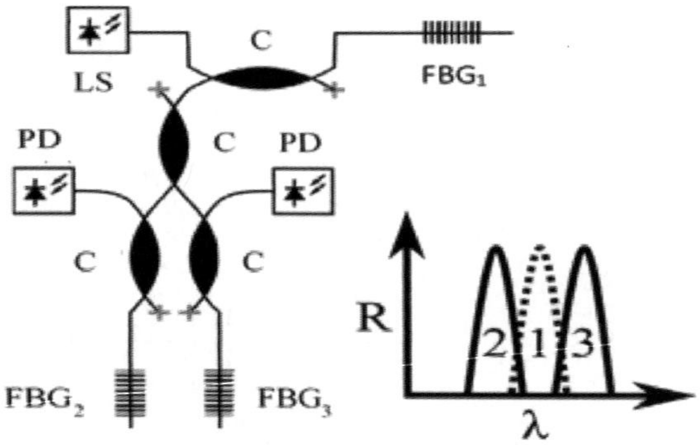

Figure 4. FBG interrogation scheme, based on power comparison method:
LS – laser source;
C – coupler;
PD – photodiode;
FBG_1 – FBG sensor;
FBG_2 – FBG_3 - reference FBGs.

The device consists of a laser source (broadband laser diode), a set of Bragg gratings (sensing FBG and two reference FBGs), fiber splitters 50/50 (couplers) and photodiodes. Optical radiation from the source, through the optical fiber and the splitter, is fed to the sensing FBG. The radiation reflected from it is applied to a pair of reference FBGs, which, in turn, reflect a portion of the incident radiation on the photodiodes. Reference FBGs are thermally stabilized in order to avoid spectrum shifts. All elements of the device are made in fiber design, which increases the mechanical stability and doesn`t require additional adjustments.

Earlier (Figure 1) we showed that when the FBG`s length decreases, the reflection coefficient decreases and the spectrum broadens (Figure 5):

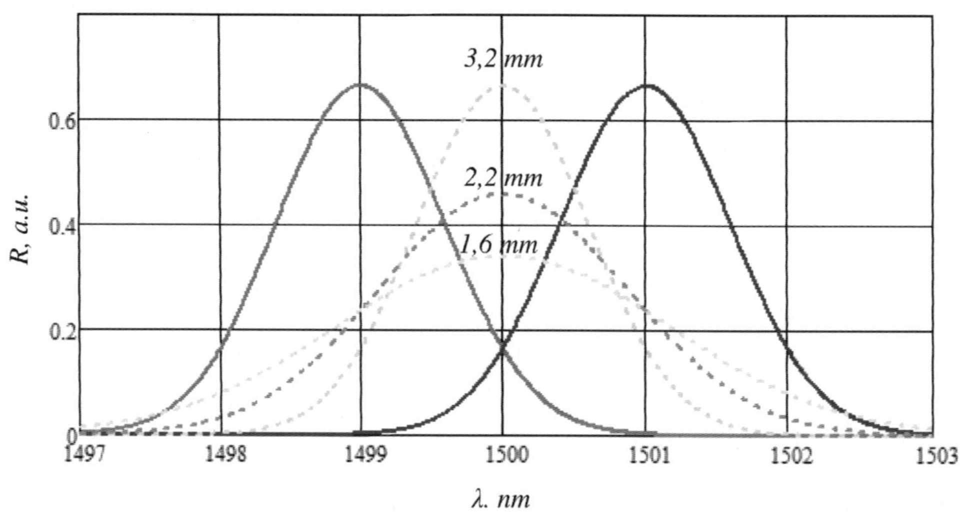

Figure 5. Modeling the operation of the interrogation scheme.

We use the total signal from two channels (reference FBGs: red and blue spectrums on Figure 5) *S2* + *S3* as shown on Figure 6.

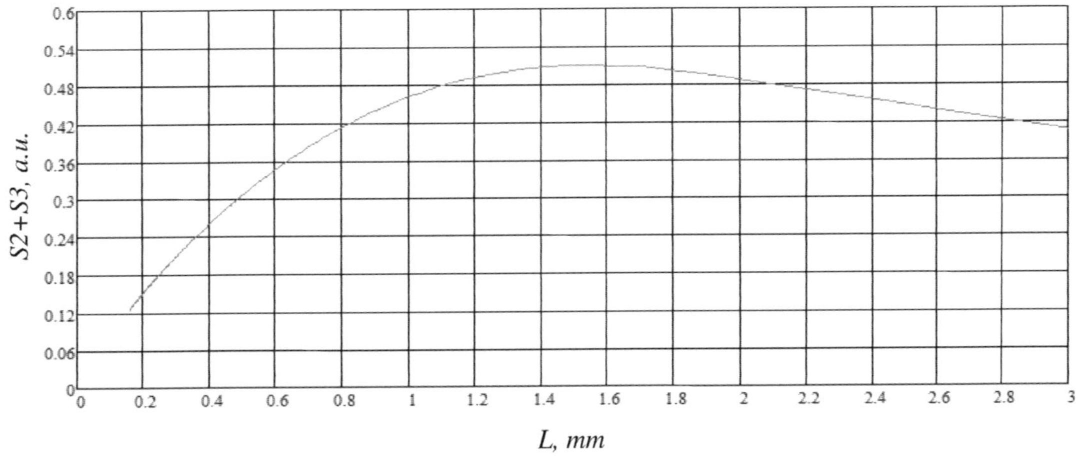

Figure 6. Measuring characteristic.

It can be seen that the characteristic has a non-monotonous section. However, taking into account the fact that wear of the brush during its operation changes monotonously, and the work of the measuring system assumes continuous measurement of its length, this circumstance can be eliminated.

It is obvious that a wavelength shift due to temperature changes of the sensing FBG will affect the measuring characteristic. To estimate the introduced error, we simulated the measurement characteristics at different positions (temperatures) of the FBG (Figure 7) over the entire range of operating temperature *T* of the sensor.

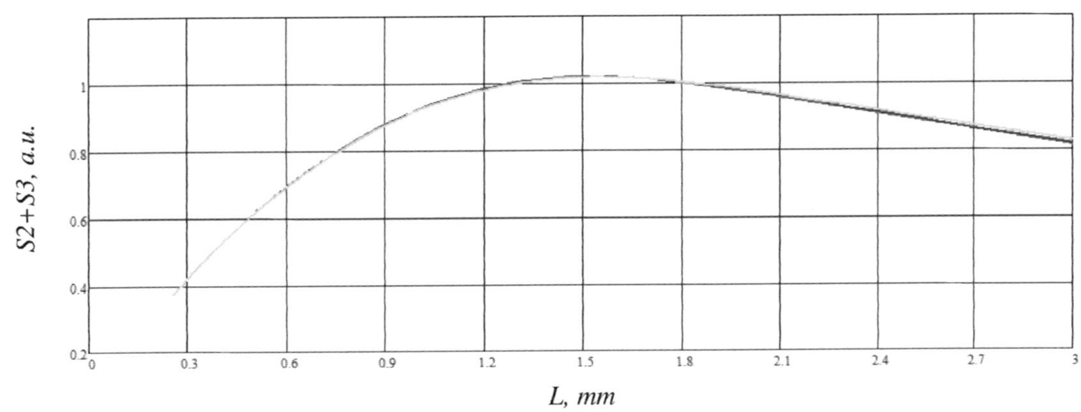

Figure 7. A number of measurement characteristics for $T = -60 .. + 120\ °C$.

The main error is caused by the change in temperature at the initial section (with the maximum length of the FBG). Based on the steepness of the characteristics in this area (Figure 7), the amount of the introduced error is less than 2% (0.11 mm). That is connected, including the small slope of the measuring characteristic in this section.

IOP Conf. Series: Journal of Physics: Conf. Series **999** (2018) 012017 doi:10.1088/1742-6596/999/1/012017

5. Experimental results

For the experimental research, we fabricated a brush with integrated FBG sensor (Figure 8). Two fibers were placed in the brush: the main and backup.

(a) (b)

Figure 8. Scheme of placement of sensors (a) and the appearance of the brush with embedded FBG sensor (b).

The testing method that we used in experiment is based on the procedure for testing brushes of electric machines in accordance with state standarts. According to the standard, brush length measurements are made by a precision tool before and after a test cycle of 50 hours. The test cycle is the continuous operation of the engine under investigation in nominal regime with brushes placed in it.

Figure 9 shows the measurement characteristic obtained during the experiment (orange line) and the calculated (blue line). An experimental study showed that the absolute error in brush length (wear) measurement was 0.11 mm.

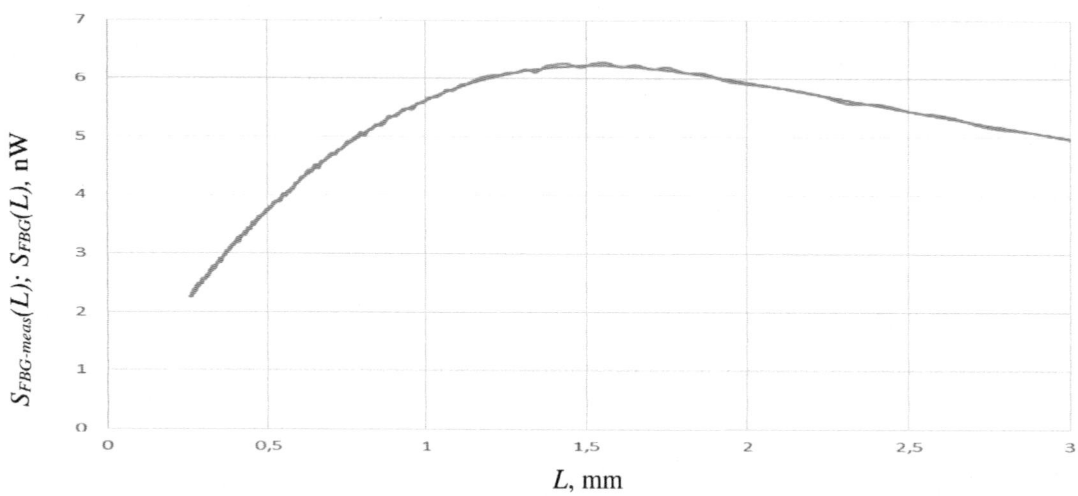

Figure 9. The dependence of the measured $S_{FBG\text{-}meas}$ (L) – the orange line – and the theoretically obtained S_{FBG} (L) – the blue line.

6. Conclusion

A new smart photonic method for characterization of carbon brush wear is presented. It is based on the usage of spectrum variations, caused by sensor`s length reduction. Two types of sensors can be widely used: FBG (for measurement range ~3-12 mm) and FBG arrays (many tens of millimeters quasi-continuous measurement). The results of development and research of mathematical models of the proposed sensors are presented. The use of the power comparison method for interrogating of developed FBG sensor is proposed. The data obtained in the course of simulations was verified experimentally with high accuracy, which confirms the correctness of the proposed solution.

Prospectively three parameters can be used for complex brush and motor: the change of FBG`s length can be used for measurement of length of the brush and it`s wear rate, grating`s central wavelength shift (due to refractive index change) for temperature regime monitoring and wavelength shift due to mechanical stress (caused by grating`s period variations) for vibration and rotation speed control.
We are first from our point of view who shown how to measure brush wear by means of FBG length monitoring and use of other FBG multiplicative properties to measure above mentioned characteristics.

The clearest example of the sensor usage is electric machine brush-collector node. But it also can be used in car brake pads, bearings, another types of friction surfaces.

7. Acknowledgements

This work was supported by Ministry of Education and Science of Russian Federation (8.6872.2017/BCh, program “Asymmetry”).

8. References

[1] Johnston R.T. 2009 Fiber optic sensor system for detecting surface wear US Patent 8571813, 1–12
[2] Sabol S.M, Subramanian R. 2007 Wear monitoring system with embedded conductors US Patent 7270890, 1–16.
[3] Pavlutskij, N.N. 2003 Device for inspection of wear of brush RU Patent 1809481, 1–14.
[4] Takeshi Satake, Yoshiyuki Imada 1989 Wear sensor US Patent 4884434, 1–10.
[5] Discenzo F.M. 2009 System for monitoring bearing wear US Patent 7551288, 40–61.
[6] Kuznetsov A.A. 2016 Smart photonic carbon brush *Proc. of SPIE* **V. 9807** 98070M-1-7
[7] Srimannarayana K. 2008 Fiber Bragg grating and long period grating sensor for simultaneous measurements and discrimination of strain and temperature effects *Optica Applicata* **XXXVIII** 601-609.
[8] Kuznetsov A.A. 2016 Complexed wear and temperature fiber-optic sensors for friction surfaces *NTVP* **1** 45-48
[9] Chan C. C. 2000 Intrinsic crosstalk analysis of a serial TDM FBG sensor array by using a tunable laser *Proc. LEOS* **vol. 40** 2643–2654
[10] Cavaleiro P. M. 1999 Simultaneous measurement of strain and temperature using Bragg gratings written in germanosilicate and Boron-codoped germanosilicate fibers *IEEE Photonics Technology Letters* **11(12)** 1635–1637
[11] James S. W. 1996 Simultaneous independent temperature and strain measurement using in-fiber Bragg grating sensors *Electronics Letters* **32(12)** 1133–1134.
[12] Morozov O.G. 2008 Methodology of symmetric double frequency reflectometry for selective fiber optic structures *Proc. of SPIE* **7026** 70260I-8.
[13] Morozov O.G. 2008 Metrological aspects of symmetric double frequency and multi frequency reflectometry for fiber Bragg structures *Proc. of SPIE* **7026** 70260J-6.

Institute of Physics
Dirac House, Temple Back
Bristol BS1 6BE UK

ISSN: 1742-6588
ISBN 978-1-5108-6126-8

CURRAN ASSOCIATES INC.
proceedings.com

International Symposium on Bioinformatics, Chemometrics and Metabolomics 2016

Journal of Physics: Conference Series Volume 835

Bogor, Indonesia
18 October 2016

IOP Institute of Physics